DAMASCUS AND PATTERN-WELDED STEELS
Forging blades since the iron age

Madeleine Durand-Charre

This new book is an updated translation from French of "Les aciers damassés. Du fer primitif aux aciers modernes" (Presses des Mines, 2007).

Books by the same author:

"The microstructure of superalloys" (Gordon and Breach, 1997)

"La microstructure des aciers et des fontes" (SIRPE, 2003 and EDP, 2012)

"Microstructure of steels and cast irons" (Springer, 2004, translated from French version here above)

Printed in France

ISBN: 978-2-7598-1173-1

This work is subject to copyright. All rights are reserved, whether the whole or part of the material is concerned, specifically the rights of translation, reprinting, re-use of illustrations, recitation, broad-casting, reproduction on microfilms or in other ways, and storage in data bank. Duplication of this publication or parts thereof is only permitted under the provisions of the French Copyright law of March 11, 1957. Violations fall under the prosecution act of the French Copyright law.

© EDP Sciences 2014

Table of contents

Introduction III
Aknowledgments V

First part :
The Blacksmith's steel spanning four millennia

1 Primitive iron
   1.1 Iron before the Iron Age 3
   1.2 Early iron making techniques 7
   1.3 The mythical aspects of iron 12
   1.4 Archaeological remains 15

2 Blacksmith steel before the Christian era
   2.1 The swords/daggers of Loristan 19
   2.2 The Celtic sword-making tradition 22
   2.3 The Etruscan swords 31

3 The swords of the first millennium AD
   3.1 Merovingian (481-751) and
      Carolingian (after 751) swords 33
   3.2 The Vikings swords 40
   3.3 Swords in China 45
   3.4 Swords forged with crucible steel 45

4 Did you say Damascus steel, Damascene or damask ?
   4.1 The debate on the origin of the naming 47
   4.2 From laminated to figurative blades 48
   4.3 Crystallization steel or pattern-welded steel 51
   4.4 Damaskeen, damaskeened, damascene 53

5 From swords to knives
   5.1 The lake of Paladru 55
   5.2 The knives at the beginning of the millennium, in the year one thousand 57

6 The swords of the second millennium
   6.1 The use of the sword in Europe 63
   6.2 Damask rapiers 65
   6.3 The engraved steel imitates and competes with the damasked steel 68
   6.4 Making gun barrels with damask 69
   6.5 The pattern-welded oriental swords 73
   6.6 Swords made with Damascus steel 78
   6.7 Japanese swords 83

7 Art and Technology in the third millennium
   7.1 The Renaissance of an art and craft 91
   7.2 The search for the best cutting edge 92

7.3 Multilayer steels 98
  7.4 Inserts and mosaic patterns 105
  7.5 Sintered steels 110
  7.6 The colored blades 115
  7.7 The contemporary wootz 119

Second part:
Formation of the damask pattern

8 Understanding steels
  8.1 Phases and phase diagrams 127
  8.2 Austenite transformation in the Fe-C system 132
  8.3 Kinetics of the austenite transformation 136
  8.4 Heat treatments 140
  8.5 Solidification structure 143
  8.6 Dendritic segregation 145
  8.7 Steels used for cutlery 148
  8.8 Optimizing microstructure 151
  8.9 Coloration of stainless steels 154
  8.10 Powder metallurgy 156

9 Pattern-welding
  9.1 Welding different layers 159

10 Moire pattern in wootz type, high carbon steels
  10.1 Crucible steels (wootz, pulad) 169
  10.2 Formation of the moire pattern 170
  10.3 Structure of the matrix 180

11 Alignments in medium carbon steels
  11.1 A well-known phenomenon 189
  11.2 Occurrence of banding in ancient steels 193
  11.3 The contribution of structural metallurgy 199

12 References
  List of references 203

13 Index
  Index 211

Introduction

Trying to understand what a damask steel is, I discovered a fascinating subject, rich in multiple facets which initially appears simple: this is laminated steel, a composite material artistically exploited. In fact, the name is confusing when considering the words refering to Damascus such as damask, damascene and Damascus steels have different meanings and refer to very different materials whose common feature is just a wavy pattern.

During the first millennium BC the Celtic smiths acquired control of hot iron working, in particular for the production of strong sword blades. The latter were built by forge welding more or less carburized pieces of smelted iron. Damask steel appears as a pattern-welded composite resulting from this know-how several centuries later.

Meanwhile, in Eastern countries a high carbon steel was developped. This legendary steel displayed a moire pattern after specific forging. It was known by several names, Indian steel, Damascus steel or wootz etc. and remained rather mysterious until the 17th century. The history of the debate comparing the different kinds of steels is evoked in chapter 4.

Since this period, modern sophisticated tools became available to observe the specimens. Many researchers, in particular the Verhoeven team, investigated the microstructure of ancient blades, thus enabling clarification of how they were forged. This is the point of view that I plan to develop in detail.

What readers are likely to be interested by this subject?

- Metallurgists, I thought about my colleagues because I regretted, after having taught the structural metallurgy during many years, not having opened a very small window, a few minutes in my timetable to explain to the students how the metallurgy of iron has developed.

- Archaeologists, the curators of a museum and all the researchers confronted with the problems of expertises of the vestiges. They can, therefore, be interested by a detailed approach to the micrographic aspects.

- Blacksmiths who have a good technical approach to metallurgy supported by sound experimental know-how. They participate on the Internet in numerous discussion forums. As an occasional observer, I noted some points for which a fundamental approach could be useful.

- Collectors and amateurs who appreciate to be informed of the making of their valuable artistic blades.

However, when writing this book it appeared difficult to me to address such different readers, to avoid specialist's vocabulary and find a common language. That is why the text has been divided into two parts. A first descriptive part which presents the developments in the technology of forging in various places and various periods, and a more fundamental second part with the aim of explaining the scientific mechanisms and present the most recent findings for the formation of particular microstructures such as the bands or alignments responsible for the moire structure.

However, steels are a material with multiple and complicated transformations; this is true even for steels of basic cutlery industry. The explanation of a mechanism requires a detailed description in order to respect a strict argumentation. To ensure a sound basis to the discussion in its scientific context, the concepts of phase, grain, segregation and the particular features of high carbon steels are revisited in chapter 8. This edition is supplemented with a few new examples, with tables summarizing the findings and an index. Ultimately, this book does not propose practical solutions to the blacksmiths but rather a microscopic vision of their metal in order to support a better understanding of the formation of the microstructure.

Madeleine Durand-Charre

mdc.damas@gmail.com

Aknowledgements

To acheive this work I ventured a little imprudently out of my strict scientific discipline. Also it was necessary for me to ask for the assistance of specialists in other disciplines extremely far away from mine. I express my gratitude to all those who believed in this project and who allowed the use of documents or samples, or prepared specific photographs.

I will quote in particular MM Paul Merluzzo, Louis Bonamour, Eric Perrin and Ernst Kläy. Many archaeologists, researchers, curators of a museum, transmitted documents or information to me: Mmes Christine Bouclet-Riquier, Véronique Despine, Anna Feuerbach, M.C. Lebascle, Aurélie R. v.Bieberstein, Bernadette Schnitzler, C. Vigouroux, et MM Gilles Desplanque, M. Ferry, Leon Kapp, J. Parisot, Jean Renaud, O. Renaudeau, J. P. Sage, Philippe Schaffnit, Pierre Thomas, Eric Verdel, John Verhoeven and Yoshindo Yoshihara.

My project began with my professional retirement when I had time available but I did not have any more the daily contact with my former laboratory at the Polytechnic Institute of Grenoble (INPG). Micrographs coming from my work of research, almost the memories of a photo album, were not sufficient. I am particularly indebted to my former colleagues and friends Annie Antoni, Florence Robault, Catherine Tassin and Muriel Véron for fruitful discussions and their assistance to supplement some examinations, analyses or micrographs.

I thank the blacksmiths, the knifemakers, the engineers and François-Xavier Salle, chief editor of "La passion des couteaux" who enabled me to give an outline of the original achievements of the contemporary artistic crafts: Olivier Bertrand, Per Bilgren, Alain et Joris Chomillier, Des Horn, Sébastien Masson, Matthieu Petitjean, Eric Plazen, François Pitaud, Denis Pittet, Pierre Reverdy, Manfred Sachse, Henri Viallon and Achim Wirtz.

I thank Jean Giraud (Moebius) for the permission to use a drawing, François Chabanne for the access to ancient books, as well as Micheline Mosselmans and Sonia Durand for their contributions to the drawings.

Avril 2014
Madeleine Durand-Charre

First part :
The Blacksmith's steel spanning four millennia

1 *Primitive iron*

1.1 Iron before the Iron Age

The metallurgy of iron appeared in most ancient societies subsequent to that of gold and that of copper. The most significant appearance was situated in Anatolia in the Hittites, between 1400 and 1700 BC, thus determines the beginning of the Iron Age. The oldest discoveries date back to the prehistoric period around 5000 years BC in Iraq (Samara), in Iran (Tepe Sialk) and in Egypt (El Gerseh). More recently, discoveries in the period known as the bronze age (3000–1600 BC) are all situated on a wide border East and Southeast of the Mediterranean Basin in Mesopotamia, Turkey, Egypt and Cyprus (more details can be found in [1]).

Furthermore, the presence of objects made from iron does not necessarily imply the ability to extract the metal from its ores, since iron also exists in native and particularly meteoritic forms, although the sources are by no means abundant.

Primitive iron

The earliest iron used by man was generally meteoritic in origin. It is the presence of nickel which distinguishes it from the other categories of iron. Modern characterization techniques enabled detecting that nickel is present in most objects of the prehistoric period and in those from the early and middle bronze ages. The iron found as metallic meteorites, called also siderites (Figure 1.1.1), were worked in the same way as stone in order to form tools. In Greenland, three meteorites among the largest ever found (one weighed 36 tonnes) had been used for generations by Eskimos.
In Central and South America, the Aztecs, Mayas and Incas used meteoritic iron well before knowing its metallurgy. They considered it

Figure 1.1.1 :
Polished section of a metallic meteorite from a multiple fall, known under the name of Gibbeon (Namibia). To notice the Widmanstätten structure consisting in long needles imbricated along three directions.
Courtesy ENS Lyon, Fr.

as extremely precious and restricted its use to jewellery and religious objects. In Egypt, the blade of a magnificent ceremonial dagger found in Tutankhamen's tomb (1350 BC) was identified as being made from meteoritic iron. It was one of a pair of objects, the other being gold.

Meteoritic iron is an alloy which generally contains a few percent of nickel, with amounts ranging from 5 to 26%, together with small amounts of cobalt (up to 1%) and traces of sulfur, phosphorus and carbon. Metallic meteorites are relatively malleable. They are one of the three major classes of meteorites, corresponding to metallic, stoney and mixed structures [2]. They are generally believed to be fragments of planets that have disintegrated, the metallic meteorites emanating from deep inner layers. The crystalline phases present in metallic meteorites have names specific to this field of study. For low nickel concentrations, the body-centered cubic crystal structure is known as kamacite (α ferrite in steels, §8.1), whereas the face-centered cubic structure found in high nickel meteorites is called tænite (γ austenite in steels). This structure consists of plate-like ferrite and was observed for the first time in 1808 in a meteorite by Aloïs von Beckh Widmanstätten (Figure 1.1.1). The plates are oriented in directions which form an octahedron. The origin of this structure in meteorites has been suggested to be associated with the existence of a solid state reaction at very high pressures [3]. However, for certain meteorites the microstructure is so coarse, with broad plates several millimeters, that a solid state transformation could seems unlikely [4].

Telluric iron may be found in the native state in basalt or other rocks as small grains or nodules. It often contains a high content of nickel, up to 70%. This iron, rarer than the meteoritic iron, was sometimes found in precious objects. Ural native iron contains 50 % of platinum.

The name *terrestrial iron* is given to iron extracted from ores ; it is normally free from nickel. Iron of this type has been found in objects in Egypt, in the Temple valley and Cheops' pyramid at Giza (2500 BC) and at Abydos (2200 BC). However, the number of such objects is small and their authenticity is doubtful, due to their poor state of conservation.

The oldest iron not of meteoritic or native origin is found as small decorative inlays in gold jewellery or tiny cult objects. It has been suggested that this iron is a by-product of the gold production process.

Magnetite is frequently present in the gold-bearing sands in Nubia and could have been reduced during the smelting operation as pasty iron floating in the slag above the molten gold. Another possibility is that iron oxides were deliberately associated with other oxides used as fluxes for the manufacture of bronze. The iron dating from this period was described as *accidental* iron [5].

Whether the production of iron by the reduction of ores was discovered at an early stage, before 2000 BC is a subject of controversy. All the more, the presence of non-meteoritic iron objects is not always associated with evidence of local mining activities. For example, in Egypt, where iron ore deposits are abundant, there is no sign of their exploitation. The argument is that it is probably due to the absence of forests capable of supplying the charcoal necessary for reduction.

Recent investigations with modern means of analysis have evidenced that a bead from the prehistoric Gerzeh cemetery, approximately 3300 BC, which was considered as the earliest example of exploitation of iron in Egypt, was in fact made out of cold-worked meteoritic iron [6].

It must be remarked that several millennia elapsed between the first reliable identifications of iron artifacts and the start of what can be genuinely termed the iron age. Several explanations can be suggested. The most obvious one is the inherent difficulty of extracting iron from its ores. The processes used for gold and copper are not applicable, and in particular, much higher temperatures are required.

It is difficult to forget what we know about iron to imagine how this new material was considered. In fact, the iron obtained by the most primitive processes of reduction of the ore should not have been regarded as an interesting material. When iron is reduced, it is pure and highly malleable, thus usable only for ornaments. It was rare, and therefore very precious and its value could exceed several tens of times that of gold.

1.2 Early iron making techniques

Iron ores

After aluminum, iron is the second most abundant metal in the Earth's crust. The major iron ores are essentially oxides (magnetite Fe_2O_3, hematite Fe_3O_4 and limonite), carbonates (siderite) and sulphides (pyrite). The preparation by washing and crushing of the ore is the same one as that practiced for the other ores. Many ore deposits occur in the eastern Mediterranean basin and can often be readily recognized due to the associated rust-red coloration of the earth. Indeed, they were often exploited as pigments, giving the yellows, ochres, browns and reds used by the Egyptians. Evidence of early mining activities is visible in deposits in Syria and Cappadocia, which appear to have been the first to be exploited on a large scale. Metallurgical culture is extremely ancient throughout the fertile crescent, facilitated by the presence of numerous rich ore deposits. The Assyrians seem to have practiced the reduction of iron ore as early as the 19th century BC.

Some ores were famous, probably due to the natural presence of alloying elements such as manganese (Siegerland in Germany), nickel (Greek or Corsican ores) or phosphorus (Lorraine ore) [7-8].

Iron smelting

In the earliest iron processes, washed and crushed ore was heated with charcoal in a primitive furnace, often consisting of little more than a hole in the ground. The temperature attained was insufficient to achieve melting and the oxide was reduced by the carbon in the solid state, leading to a spongy agglomerate called a bloom. Many primitive furnaces were built in such a way as to optimize natural drought (Sri Lanka). Furthermore, the use of rudimentary bellows made from animal hide was probably adopted at an early stage. The furnace can be controlled by the injection of air and, depending on the setting, local carburization of the iron can be acheived. This carburization occurs at high temperature by contact of the iron with a CO atmosphere near the charcoal. Traces of cast iron found amongst the slag in ancient smelting centers indicates that the temperatures attained were sufficiently high to induce partial melting. However, such cast iron was probably initially obtained accidentally and considered as a worthless by-product, since it was hard, brittle and unworkable.

The development of iron smelting was particularly facilitated in areas where ore deposits were associated with ready supplies of charcoal and refractory materials for furnace construction.

The smelting stage produces a spongy agglomerate called a bloom, consisting of well reduced iron particles, residual oxide inclusions and areas with carburized iron particles. The bloom has to be repeatedly heated and hammered to expel residual slag inclusions, forming a more compact mass. The addition of sand during reheating contributes to the removal of oxide inclusions by the formation of a fusible envelope of fayalite. The iron obtained in this way was fairly pure, since oxides of other metallic elements such as Si or Al could not be reduced in these conditions. Its carbon content is low, it is therefore malleable.

Pure iron can be carburized by simple welding in contact with more carburized iron agglomerates during repeated hot hammering. If the metal working is intense, the agglomerates are crushed, mixed and carbon diffuses between the iron grains, it results in a homogenization of its concentration.

The soft iron can also be carburized by a further treatment known as cementation whereby carbon difffuses into the metal. At the temperatures attained, the depth of carbon penetration was no more than about a millimeter (see Table 8.1.2). Thus, this process can be carried out only locally to strengthen superficial areas, or sharp edges, or on iron divided into pellets or thin strips. Prakash underlines that the Indian blacksmiths had acquired a great knowledge of the effects of carbon in iron and knew how to control the production of more or less carburized iron [9].

Crucible steel in East Asia

The fusion of steel in a crucible is an invention of the first millennium BC. The production of *crucible steel* goes back to 300 BC according to the history of the wootz written by Srinivasan and Ranganathan at the University of Bangalore in India, [10-11]. This date is based on the discovery of high-carbon steel on a production site of this period (Kodumanal). This vestige is the oldest known, it is presented as a relatively reliable proof of the more or less complete fusion of the steel.

The same authors also quote a document according to which, Indian king Pôros, defeated by Alexander Great, offered to his conqueror 30 pounds of Indian iron in 326 BC. Several indices let suppose that it was most probably crucible steel. However this assumption remains rather controversial.

Recent excavations revealed several sites of production in Central Asia, among them Merv in Turkmenistan and Achsiket in Uzbekistan. The oasis of Merv located on the silk route was an important production center in the 8th et 9th centuries. Anna Feuerbach, who took part in excavations, presents in several publications an example of exploitation of vestiges led in a rigorous scientific way [12-15]. The sites of Achsiket are dated between the 9th and 14th centuries.

The iron as reduced from its ore was melted in a crucible in a distinct subsequent operation. Steel was prepared by small charges, each consisting mainly of iron, charcoal of bamboo and leaves of specific plants. The iron used was probably in the form of an iron sponge, it still contained oxides particles which while melting form a protective slag at the surface of the ingot. The whole was then hermetically sealed in clay crucibles. A batch of some twenty crucibles was heated for a long time at temperatures up to 1200°C, and then let cool. Liquid iron was carburized by reaction with charcoal enabling a partial melting [12-13].

The iron mass found in the bottom of the crucible was a small ingot called *cake*, typically weighing up to 2 kg [9]. Iron prepared in this way differed from other irons by its high carbon content, up to about 1.5 %. Indeed, it is a high carbon steel. Trace elements such as vanadium and titanium, possibly from the bamboo charcoal or other plants employed, probably contributed to the exceptional properties of this steel (§ 10.1).

The metallurgical know-how of India was highly appraised as early as the Pre-Islamic period [11]. The method of preparation by fusion, initially exploited in the southern part of the Indian sub-continent including Sri-Lanka, slowly propagated originally towards Asia and then towards the Middle-East, Iran, Turkey and even as far as Russia. Later, this steel will be known under the name of *wootz*, a name which undoubtedly comes from the English deformation of the ancient Indian word Ukko and which was stabilized at the beginning of the 19th century [16-17]. Unfortunately vestiges of this time are very rare. According to Srinivasan and Ranganathan [11], one of the reasons is perhaps that the Indians had, at least during the historic period, no funerary tradition of tombs like many other civilizations.

Thus the period during which this process was used extends from 300 BC to 1856 AD. This last date corresponds to the beginning of the modern age of iron in 1856 with the invention of the Bessemer process to manufacture molten steel on an industrial scale, which cannot be compared with the manufacturing of the Indian steel cakes. The fusion of steel was invented by Benjamin Hunstman in Sheffield in 1740, approximately two millennia after crucible steel.

Cast iron in China

Iron production by smelting appears to have been known in China from about 1000 years BC. However, China was the first country to use cast iron *(i.e.* molten) in around the 6th and 5th centuries BC [1, 18-21].

Among the vestiges which support this discovery are cast iron cauldrons dating from 512 BC and cast iron molds from the end of the first millennium BC. It has been suggested that these developments were made easier by the presence of phosphorus-rich ores, since phosphorus lowers the melting point of the carbon-iron alloy. Furthermore, technical know-how in related fields was more advanced in China than in other parts of the world. For example, in the case of pottery, the Chinese mastered the manufacture of both red pottery, baked in oxidizing atmospheres, and black and egg-shell pottery baked in reducing environments. Their furnaces were ingeniously designed and made from high quality clay refractories, and bellows were in regular use in the 4th century BC. Their superiority was maintained by improvements such as the introduction of piston bellows in the 2nd century BC, and the replacement of charcoal by coal in the 3rd century BC, nearly two thousand years before Europe.

Several original techniques were developed to transform white cast iron into iron and steel by decarburizing heat treatment and also under the combined decarburizing action of forging and heating. Under the Han dynasty, in the 2nd century BC, cast iron was decarburized to transform it into ductile iron. Early in the 5th century AD, an original iron carburizing technique was developed, consisting in immersing rough iron in cast iron and then subjecting the coated product to a series of forging and folding cycles to produce a homogeneous medium carbon metal. Even more surprising is the recent discovery of cast iron objects dating from the Han and Wei dynasties (206 BC to 225 AD) containing graphite nodules similar to modern spheroidal graphite iron, invented in... 1948! Chemical analysis revealed none of the nucleation agents used today for spheroidization. It has been suggested that an appropriate Mn/S ratio enables graphitization of cementite to occur in the solid state with a nodular morphology [22-23].

Iron making in Africa

In Equatorial Africa, neolithic practices were directly followed by an iron age, with no intermediate use of copper or bronze. The analysis and dating of many small artifacts indicates that the metallurgy of iron in this area goes back to at least the 3rd millennium BC, and possibly even to the 4th [24]. In Gabon, furnaces dug into the soil have been carbon dated to the 7th century BC, based on charcoal residues found in the vicinity. However, the lack of spatial coincidence does not provide unambiguous proof. Furthermore, iron objects remain relatively rare, due to the difficulty of conservation in the prevailing moist climate and acidic soils.

It has been suggested that liquid iron was obtained at an early stage in prehistoric furnaces found near Lake Victoria in Tanzania, high temperatures being attained by the injection of preheated air. However, what is probably more important is that, in the region concerned, the iron ore is extremely rich in phosphorus, while the local vegetable matter mixed with the ore is also rich in phosphorus, facilitating melting.

From Asia Minor to Europe

Around 1500 BC iron smelting was practiced in the Caucasus region Northeast of Turkey. The metallurgical know-how propagated slowly. Towards 600 BC, the Etruscans and the Catalans practiced an advanced metallurgy. And in 300 BC, the Celtic culture spread as far as Ireland. The metallurgical culture expanded over all of Europe. The following well documented books can be used as initiation to the history of iron specifically to this period : [18, 25-29].

The competition between bronze and iron swords lasted for several centuries. The Romans, who had conquered territories in Spain containing rich deposits of copper ores, used bronze swords. They did not see the advantage of changing to iron until the Punic Wars against the Carthaginians, in the 3rd and 2nd centuries BC [30]. Moreover, several Roman authors ironically criticize the poor quality of the Gallic swords, *insufficiently wrought* that the warrior has to straighten after the first blow, provided that his enemy leaves him enough time !

1.3 The mythical aspects of iron

Iron fallen from the sky

How could primitive tribes perceive the meteorites falling from the sky? Whatever the place, they gave it a divine character. Quoting Mircea Eliade [31]: *Let us retain this first religious valorization of the aerolites : they fall on the ground loaded of celestial sacrality, they thus represent heaven. From there, probably, the worship dedicated to so many meteorites or even their identification in a divinity...*

Iron, the cursed metal

Iron of the blacksmith is born from fire. Fire that is worshiped, feared and hated. Similarly, even among evolved peoples, a magic, beneficial or malefic power is attributed to iron, which is considered almost in a religious manner. For certain people, it incarnates especially a devilish, harmful spirit, the bad eye. The fact that, unlike gold or meteoritic iron, smelted iron rusts makes it appear as an impure material, cursed by its degradation.

It is probably the reason for the multiple taboos, interdicts and superstitions. Among Egyptians, iron was regarded as one of the attributes of the god Seth as well as a russet-red hair and consequently was named the *bone of Seth*. More recently, the Israelites of king David's time showed a similar aversion, forbidding any use of iron tools to cut the stones of an altar. The classical Greeks even composed a prayer to prevent rust.

In other periods it was disadvised against using it to cut herbs and meat. In Africa, the iron tools were accused of moving away the rain from the grounds plowed with it. The literature relative to the beliefs and practices which surround iron is incredibly rich in revealing anecdotes.

The blacksmith's place in society

In the same way, the attitude towards the blacksmith was different following civilizations.

- The one who controls the fire must be venerated because he forges weapons in the imitation of the divine blacksmiths who mastered thunder and flashes of lightning to arm the Gods, he tames the divine element of fire. Nickel remarks in his introduction to *Damascus Steel* from Sachse [32] : *it is significant that the only human craft which was found sufficiently worthy by one of the Olympian gods (Hephaitos/ Vulcan) was that of the smith.*
- The one who masters the fire inspires fear because he achieves a demonic work. The rites of melting iron comprised sacrifices. Certain Babylonian, and more recently African, ancient ritual practices report collective sacrifices, which sometimes included humans or foetuses .

All over the world the primitive practice of the fusion of iron is associated with a symbolic, almost liturgical system [31]. In Africa, more than anywhere else, iron making practice was closely tied to social structures. Africa is the only continent where iron continued to be produced and worked according to ancestral customs until the middle of the 20th century [33]. Ethnologists have been able to directly question elders and even to reproduce the ritual melting of iron in the social context.

The smith was an important person, being both a craftsman and a sorcerer. He supervised preparation of the furnace and the smelting operation, and was the grand master of a ceremony celebrating the *espousal of ore and the inferno* and *fecundation by fire*, to give birth to iron. The great day of smelting was preceded by purification rites and abstinence. The feast was accompanied by music and incantations.

In a less mythical way the Celtic smiths were honored as masters. Their rich burials in the civilizations of Hallstatt and La Tene testify to this consideration.

Swords, witnesses of the know-how of the blacksmiths

The history of metallurgy can be read from the evolution of many different objects commonly found in archaeological excavations, including axes, ploughshares and pins. However, weapons, from knives to canons, have always been the first to benefit from the most recent technological progress. The bladed weapon was the primary tool of battle for three millennia. But the sword is moreover a mythical and sacred instrument. It represents an attribute of social status, a luxurious and infinitely invaluable object. Because of this symbolic role, richly decorated objects were preserved, transmitted like a heritage, and were kept as sacred objects. The popular legends tell the story of these swords: Durandal in Roland's song, Excalibur in the legend of king Arthur and many others. The legend of Wieland, taken back by the operas of Wagner, was known from the 6th century but the tradition goes back in time probably farther. The mythical aspect resulted in inscriptions, decorations, and real rites surrounding the manufacture. Forging a sword goes beyond the simple craft work, it is a fulfillment. The common phantasm is that a strength is transferred by the blacksmith and his environment during the manufacturing of the weapon.

In Muslim countries, the religious aspect is clearly illustrated by inscriptions of verses of the Koran on swords, and especially the symbolic decoration in ladder, said ladder of Muhammad or the 40 steps, to remember how Allah welcomes the valiant warrior killed during the holy war. In Catholic countries in the Middle Ages, knights had their sword blessed during the ceremony of dubbing. In Japan, messages of Buddhist or Shintō inspiration were engraved on swords of the samurai. This sword was the symbol of his honor and a kind of talisman.

To simplify, the weapons named sword, dagger, saber etc. will be often indicated under the general name of sword in the continuation of this text, without taking into account the particular appellations. A sword comprises a handle and a blade. The handle consists of a pommel, the handle itself and a guard. The pommel is supposed to balance the weight of the weapon. The guard is intended to protect the hand. The blade can be right or curved according to whether the sword (saber in the latter case) is designed to slash with the sharp edge or to thrust in the axis of the sword to pierce the opponent. The saber has a blade with only one cutting-edge while the sword has two cutting-edges and a scramasaxe is a kind of semi-long saber. Many particular names are used for these stabbing and slashing weapons which correspond to a design specific of a period or of a country.

1.4 Archaeological remains

Necropolis

Ancient residential areas contain the vestiges of the everyday life and sometimes the traces of workshops and forging furnaces. Moreover the presence in tombs of precious objects such as swords, daggers or tools reflect the reputation of the deceased craftsman and his social status.

The necropolis of Hallstatt discovered in 1824 is considered as a representative reference of a wide period. Hallstatt is the name of a village in Austria where excavations, performed from 1846 till 1963, brought to light a rich iron age cemetery with many objects dating from 1200 to 500 BC. Iron items were found in the third level at this site, called Hallstatt C, extending from 800 to 600 BC, which corresponds to the beginning of the iron age in this region [34].

Tombs of Celtic warriors were discovered in several sites in Europe, they often contained a rather complete set of fighting equipment (axes, spears, shields, helmets, daggers, swords). Swords were found there completely twisted, made deliberately unusable (Figure 1.4.1).

Figure 1.4.1 :
Bent iron sword, 91 cm long, found with burnt bones in a Gallic cemetery dating from the middle of La Tene period, around 200 BC. Numerous twisted swords were found in the tombs of Celtic warriors and appeared to have been ritually made unusable. Courtesy Musée Dauphinois, Grenoble, France.

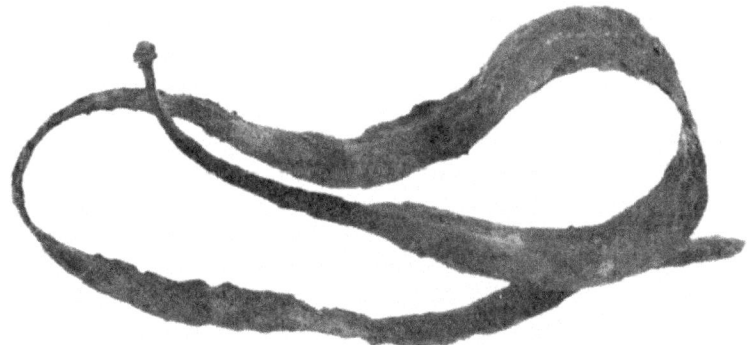

Figure 1.4.2 :
Votive offerings in river worshipped as a deity.
Courtesy Musée de Chalon sur Saône, sketch for the Journal of roman military equipment studies, [35].

River beds, lakes, peat bogs

Numerous objects were found buried in muddy deposits of rivers under silt (Saône for the Gallic objects § 2.2, Loire and Seine for Vikings objects § 3.2), buried in lakes (Paladru, for the An Mil knives § 5.1), buried in peat bogs (Denmark, objects dating from the 2nd and 3rd centuries). Sometimes these objects were in poor condition, badly damaged by oxidation, but fortunately some were found in excellent condition when they were protected by a layer of clay, or when the aqueous environment was not acid.

The second reference site for the iron age is La Tene, a river bed north of Lake Neuchâtel in Switzerland in which excavations began in 1858. This rich site dates back to between 500 and 50 BC. An artistic style was named after it. In fact, the La Tene culture is derived from that of Hallstatt, but is more homogeneous and more typically Celtic. The gold and bronze artifacts were richly and imaginatively decorated.

La Tene is not a single example since many other vestiges were found in rivers on dredging, or in graves by the riverside. In France, the localization of rich vestiges in the banks of the Saône, along strategic passages, for instance close to the piles of ancient bridges or close to a ford suggested that weapons had been given up after a battle. However, the variety, the places and the dating of all that was found (swords, helmets with coats of mail folded inside, diverse metallic bowls) suggest another explanation. These valuables could be considered as offerings to the River worshipped as a deity (Figure 1.4.1). Such pagan worships continued to be practiced for a long time after the beginning of christianization. The habit of throwing or depositing an offering in water remains nowadays although the reason for it is completely forgotten.

Restoration and examination of vestiges

Metal vestiges must be treated to restore their primitive form as well as possible and to stabilize their conservation. In the book *In search of lost metal* [36], it is recalled that one of the basic principles of the preservation-restoration is the reversibility of the treatments, *i.e.* the possibility of eliminating them to return to the state in which the object was found. Thus the metallurgist is seldom authorized to cut samples out of the swords!

The first task is to remove the gangue which envelops the objects (Figure 1.4.3A). The objects are carefully erased under a binocular magnifying glass or an optical microscope with instruments similar to those of the dentists. Micro-sandblasting can be used, adopting the sanding abrasive powder to clean the object.

The second task is to stop the corrosion. Iron vestiges are currently covered with layers of corrosion products containing anions, in particular chlorides, liable to extend the corrosive action. Protection consists in neutralization of anions. Several chemical or electrochemical processes of dechlorination are used according to the nature of the gangue which is different depending on whether the object remained in a marine or a terrestrial environment. The treatment of dechlorination is very long, it lasts several months, even for objects of small size.

Non-destructive modern means of investigation such as X-ray can highlight an internal structure, for example Figures 1.4.3 B et D. Also scanning electron microscopy makes it possible to analyze areas on a microscopic scale. In the case of blades the problem is collecting representative samples without inducing damage.

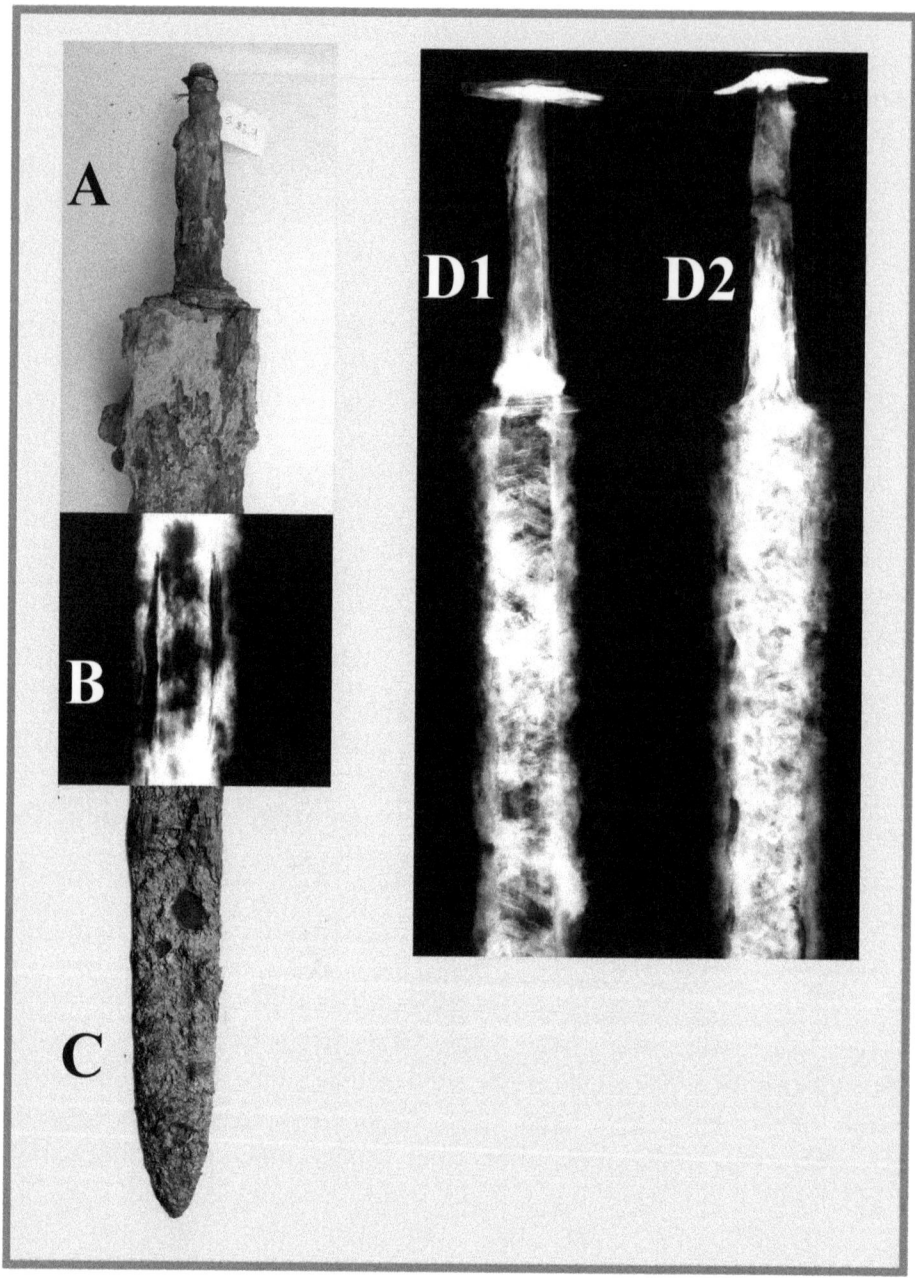

Figure 1.4.3 :
X ray examination of Merovingian swords. A) sword in its gangue; B) X ray radiograph; C) sword after restoration; D) swords radiographs.
Courtesy Laboratoire d'Archéologie des Métaux, CCSTIFM, Jarville, Fr.

2 Blacksmith steel before the Christian era

The basic know-how of the smiths was initiated during the first millennium BC.

2.1 The swords/daggers of Loristan

The daggers of Loristan are very representative of the early iron working because, to quote France-Lanord [37], it is the first example in the history of iron *of a localized production, quite typical, and distinctly industrial*. Many weapons were found in the area of Loristan (Turkey, Iran) among which appear short iron swords beside items considered as more precious made of copper, bronze or gold. They testify to a manufacturing activity located in the mountainous regions of the high plateaus. Regrettably the iron objects are often found very corroded, the place of their discovery is not precisely known and their dating is subject to controversy. According to the historical context, it seems that such weapons could be forged in the 11th century BC by the Kassites, people of horse breeders and warriors who had developed the art of metallurgy. More probably this production would be dated to the beginning of the Iron Age between the 12th and 8th BC.

Iron swords designed as bronze swords

Iron objects were found by the hundred in Loristan. Unfortunately, they were dispersed and even some probably lost for lack of restoration. The style of all these objects is homogeneous. The great originality of their manufacturing is that they were built from separate parts, held together by rivets, without welding, as the similar bronze

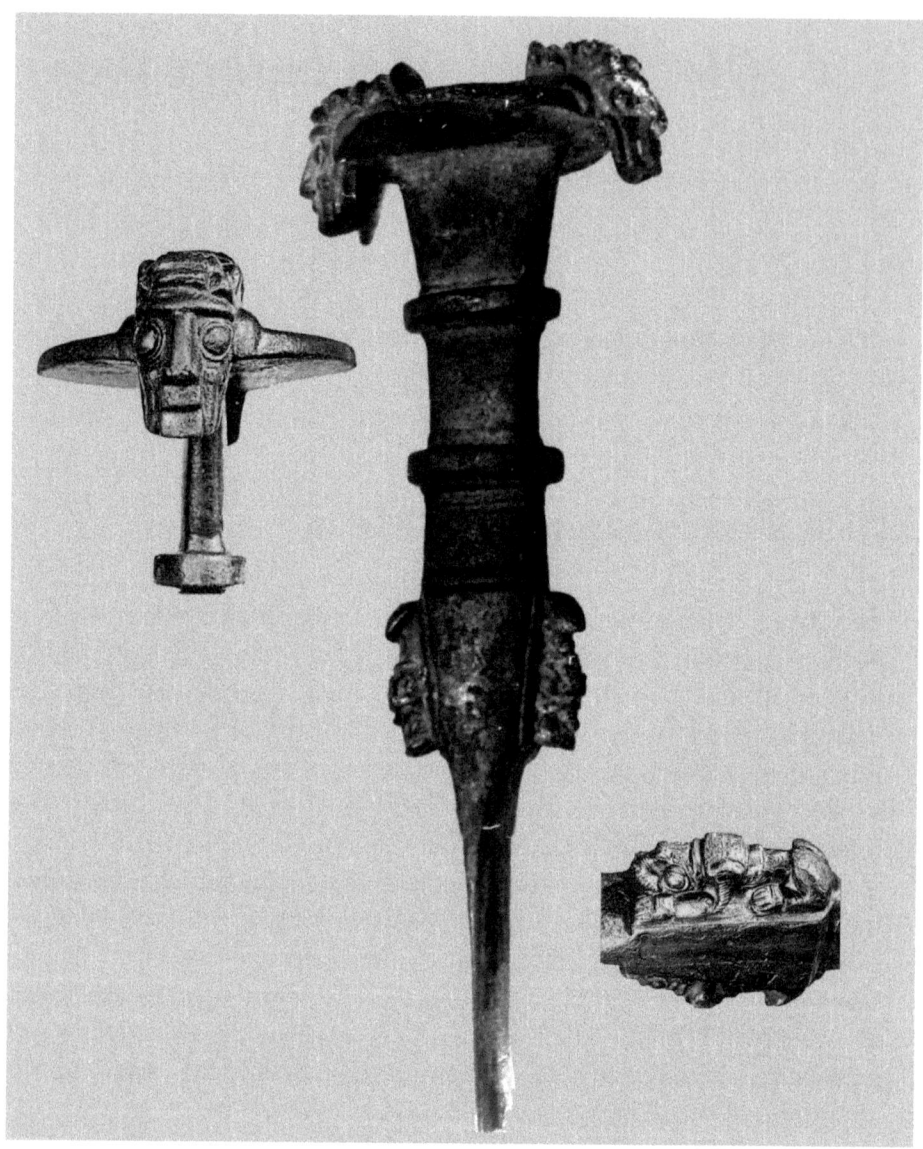

Figure 2.1.1 :
Upper part of a dagger of Loristan, overall length approximately 50 cm, dated around the 11th century BC.
The pommel has on a side a figure of a monster and on the other a human figure (magnification left). The guard is decorated with squatted lions (magnification right).
According to France-Lanord's documents, courtesy CCSTIFM, Jarville, Fr.

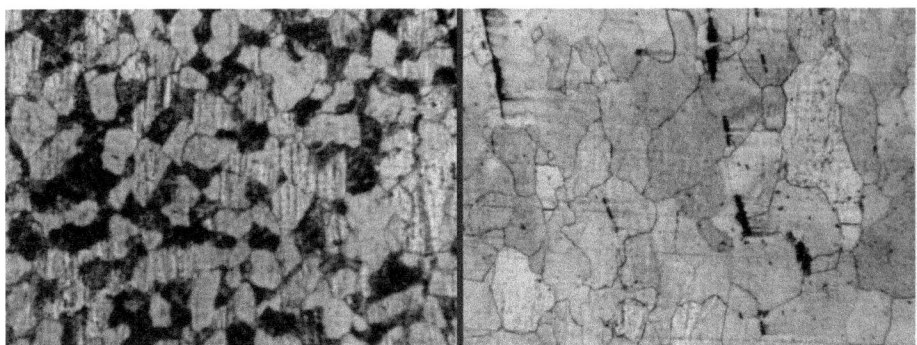

Figure 2.1.2 :
Optical micrographs of the blade on two zones of 1.3 mm wide; on the left a high carbon zone with small brown islands of pearlite cells, the proportion of which corresponds to 0.2-0.3 weight percent carbon; on the right pure iron with relatively coarse grains. The dark elongated precipitates are oxide inclusions along a weld line (see § 9.1).
Micrographs from France-Lanord, Courtesy CCSTIFM, Jarville, Fr.

parts. The decorations are not cast but rather carved parts (Figure 2.1.1).

A clever choice of more or less carburized iron

The carbon content of the steel is different depending on the parts, the choice reflecting a certain appreciation of the respective grade abilities. The blades are of better quality than the other parts, more carburized, but heterogeneous as shown by the two optical micrographs in Figure 2.1.2. This aptitude to differentiate irons revealed *a very high level of technical and intuitive knowledge of the remarkable material called steel now* [38].

Which techniques of carburization were known at that time? There is no artifact supporting any answer. An assumption is that the different high-carbon agglomerates constituting the iron bloom after reduction could be sorted according to their appearance and forged according to the future use.

2.2 The Celtic sword-making tradition

The Celts, skilled craftsmen

The development of the metallurgical practice of iron is bound to the expansion of the Celtic civilization throughout Europe from about the 6th century BC. The Celts formed a cultural unit made up of about fifty small autonomous tribes localized more particularly in the area of Central Europe between the Rhine and the Danube [34]. These tribes included well-differentiated ethnic groups within a territory. But they shared the same culture, the same basic social organization within the tribes and the same Gods.

Celtic languages of Indo-European origin were close together, but distinct from the Latin, Greek or Germanic languages of their neighbors. The Celts left few written testimonies, however they left a lot of vestiges that testify to their excellent craftsmanship, particularly for metal working. Among these vestiges were weapons which they had a great consumption of. The competitions between tribes generated perpetual conflicts. Many weapons were found according to the tradition to bury warriors killed in the battle with their arms, including enemies (Figure 1.4.1).

The excavations of the sites of Hallstatt and La Tene go back to the second half of the 19th century and new excavations came to supplement the collection of vestiges in 20th century. The latter is sufficiently rich to make it possible to outline an evolution through one long period of Celtic culture. Weapons, found by thousands, are diverse and the Celtic sword could be defined, following Pleiner [39], as a generic name referring to a weapon of a particular conception and found mainly in zones known to have been occupied by the primitive Celts. This name includes daggers (25cm) for the weapons of the first period of La Tene, then short swords (40-50cm), see Figure 2.2.1, swords dated of 3rd-2nd centuries BC and finally the swords labelled long swords (60-90cm) of the third period.

Around 600 BC a period followed during which spears were the main weapons, then again swords, the long swords reappeared. A long sword should be finer than a short sword without being brittle. Its manufacture is more difficult and represents an improved metallurgical know-how.

They made, moreover, the reputation of the Celts because the Greek and Roman soldiers at that time had only short swords. However the choice between the long sword or the short sword does not depend simply on the know-how at that period but rather on various types of engagements; the long sword being rather restricted for single fights between chiefs than for real wars between tribes. The short sword presented in Figure 2.2.1 is typical of the La Tene II period, *i.e.* almost a millennium after the period of the Loristan swords. As for the latter, the anthropomorphic hilt shows a certain degree of forging skill since it is completely forged in only one part and is welded to the blade. The blacksmiths acquired the control of welding and know-how to assemble different steels more or less perfectly. It was an essential step. The ability to carburize iron and to forge weld different metallic materials is demonstrated by certain Gallic artifacts dating from the 3rd and 2nd centuries BC [8].

Iron working

The starting point is the iron obtained from the blooms from which oxides were removed. They are hot worked in small compact blocks of about 2 kg. These blocks are then converted into heavier bi-pyramidal ingots or into bars of various forms. Such bars, found in great quantities, were a half-finished product used as trade currency.

Figure 2.2.1 :
Short Celtic sword (length 37 cm, maximum width 9.6 cm, maximum thickness 1.6 cm). Courtesy Musée d'Annecy, Fr.

Their shape and their volume are completely adapted to the manufacturing of a sword. Pleiner ordered forging of similar bars and the test was conclusive [39].

Archaeologists had anticipated this conclusion by calling certain bars salmons of sword (salmon-shaped swords) and Schwertbarren. The bars result from the welding of smaller bars, bands welded side by side or kinds of piled up plates, or the welding of very irregular pieces from the bloom. The various approaches of forging revealed by the vestiges show that the blacksmiths knew well how to work the iron from the first millennium BC (see the references [27-28, 40]).

Blades structured in stackings of longitudinal welds

Blades built by assembly can present a weakness at the level of the welds, so it is essential that the welds are longitudinal so that the future blade resists the transverse blows. It is the case of the Gallic sword in Figure 2.2.2 which appears made up of longitudinal layers. The micrographic examination of the surface of a similar, less oxidized sword, shows the alternation of more or less carburized layers. But to understand the construction of the blade it is necessary also to examine a cross section, then scientific curiosity enters into conflict with the concern for preserving an archaeological artifact intact.

In Pleiner's experience mentioned above, the bars or *salmons of sword* could be transformed into blades at the end of 16 to 36 operations of heating, followed by forging, according to the shape of the blade. To this number must be added the number of operations required to form the bar. The finished blade underwent about fifty operations approximately. The micrographs below show that the welds leave traces which remain visible even after many operations, and the blades keep the memory of the stacking mode of the initial assembly (see § 9.1).

The sharp edge hardened by carburization

The edge is the thinnest part of the blade. It was welded to the main body in the old blades, hence the name of reported edge. Some specimens among the oldest have been found with soft iron edges. This iron is malleable, little resistant, the only advantage seems to allow sharpening and easy repairs. The edges were most often composed of more carburized iron. The advantage is the possibility of a significant hardening by rapid cooling called quenching.

Figure 2.2.2 :
Gallic sword dated to the 5th or 4th found in the river Saône. The strong oxidation dug longitudinal strata. A similar sword, found on the same site, but much less oxidized revealed a microstructure formed by various layers of ferrite with a more or less large proportion of pearlite corresponding to low-carbon steels 0.2-0.3 in weight percentage.
Courtesy Musée de Chalon sur Saône, micrograph Grenoble INP, Fr.

Quenching induces a crystallographic transformation of austenite into martensite (§ 8.2) but such a structure has rarely been detected in Celtic swords. From observation of numerous micrographs published by Pleiner it appears that the blades are mainly pearlitic [39]. Why were the cutting-edges not more carburized and hardened since the role of quench hardening was well known in the first millennium? The advanced argument is that the smiths found it difficult welding different grades. This is true because the expansion induced by the transformation into martensite can cause a crack, in particular at the level of the weld. Moreover, primitive steels contained very few alloying elements due to their mode of preparation by reduction. In that case, it is the pearlite which forms most quickly, preventing the martensite formation (see § 8.4 and Figure 8.3.1).

Some rare objects of the 13th and 12th centuries BC evidence the knowledge of transformation of iron into steel and in particular quenching. For example, a famous peak of miner found in mount Adir in Galilee presented laths of slightly tempered martensite. The quenching was thus known and it was probably practiced, but the presence of martensite is difficult to detect on prehistoric objects because of the highly corroded state. In fact, martensite is more sensitive to the corrosive attack than the pure iron and it is the first one to disappear.

Typical examples of construction of a blade

Let us observe in detail three typical blades to understand their construction. A small sketch on each figure explains that the blades were sectioned in a plane perpendicular to the main axis of the blade. The optical micrographs were performed on polished surfaces slightly etched. The images are the result of meticulous assembly of several tens of micrographs for each section.

The first blade in Figure 2-2-3 is illustrated by the assembly of some twenty optical micrographs of a section of 9,5 cm length observed after polishing and etching. It is a sample of the Laboratory of Archeology of Metals, CCSTIFM at Jarville, of an undated Gallic sword. It has a traditional structure made up by welding bars side-by-side which appear in the form of strips separated by dark lines of oxides, traces of a weld. The strips comprise more or less dark zones corresponding to different proportions of pearlite. They also show lines of oxides weakly marked, less continuous. These are the traces of the preparation of the bars by repeated operations of folding, welding and forging (see also Figure 5.2.3). The reported sharp edge wraps the body of the blade so as to make the welding less liable to break.

The second blade shown in Figure 2.2.4 is made up of several welded bars but the lines of oxides are arranged in diagonal strips, thus showing that the stacking of plates was shifted during forge-welding. But the most remarkable observation is the presence of more or less broken carbide lines, parallel to the weld lines. The constituent layers are coarse grained ferrite in which the absence of pearlite indicates a very low carbon steel, perhaps even pure iron. An explanation is that the blacksmith compensated for the low carburization of iron by practicing a carburizing heat treatment before welding the layers. Heating iron in the presence of charcoal can result in superficial carburization with an appropriate control of the air flow in the furnace.

The third blade of Figure 2.2.5 is that of a typical Gallic sword with its iron sheath. The body of the blade is forged in a bar of a relatively homogeneous steel. Some thin oxide lines show that steel was worked repeatedly to obtain a remarkable homogenization of the carbon content. The blade is highly carburized and presents some martensitic areas close to the cutting-edges, more exactly, with the microstructure of a slightly tempered martensite. The rhombus shape of the section corresponds to the presence of a central ridge on each side of the blade. It is formed during forging and is intended to reinforce rigidity. Another quite Gallic characteristic is the presence of a sheath of almost pure iron, worked out of a sheet thinned in order to limit the total weight.

All these observations show that the Gallic smiths knew remarkably well the metallurgical techniques including: the preparation of steel, heat treatments and the craft to work sheet steels. Thinning of iron was probably carried out by cold hammering. Numerous similar craftwork objects were found: helmets, bowls and sheaths.

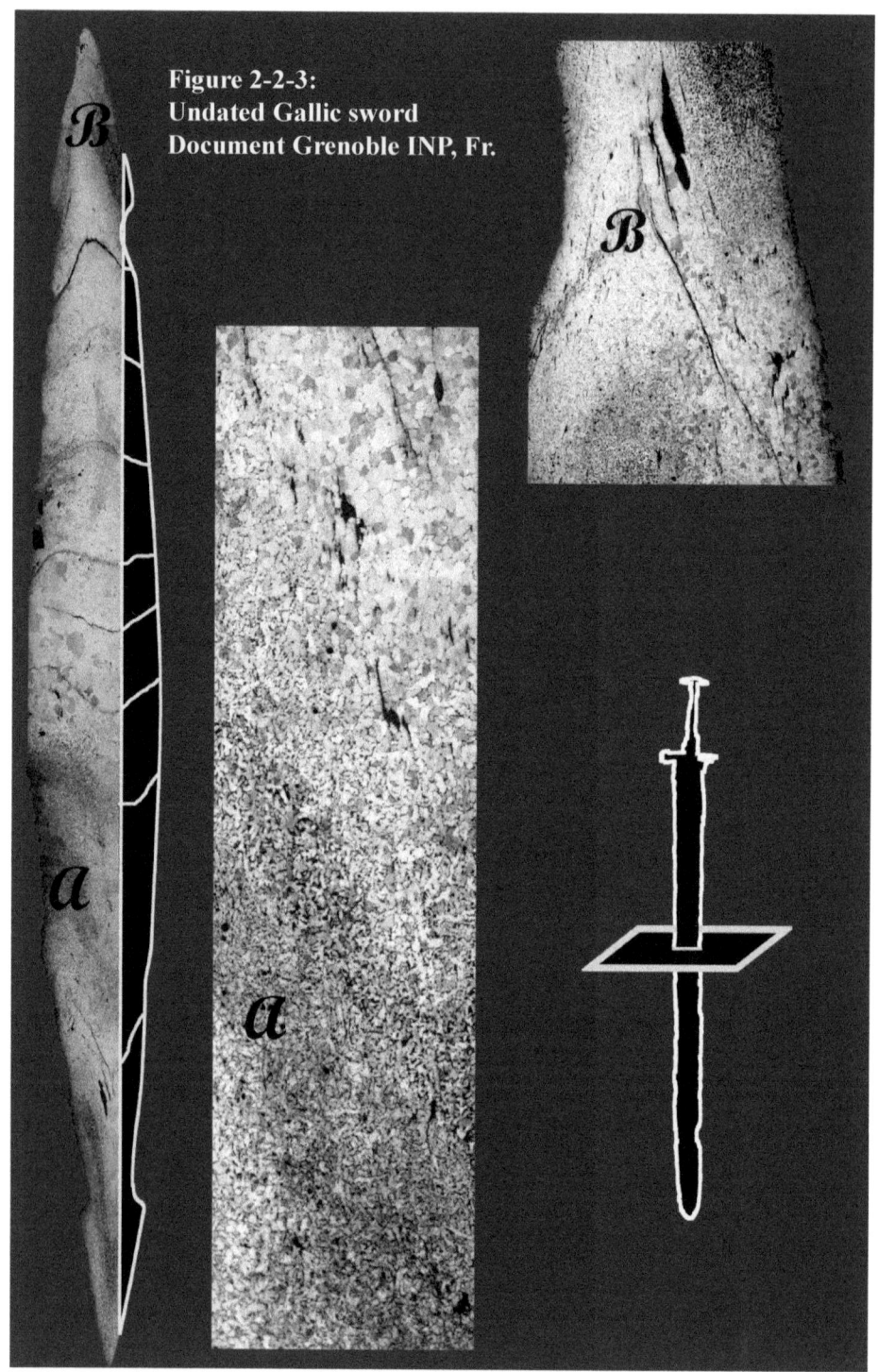

Figure 2-2-3:
Undated Gallic sword
Document Grenoble INP, Fr.

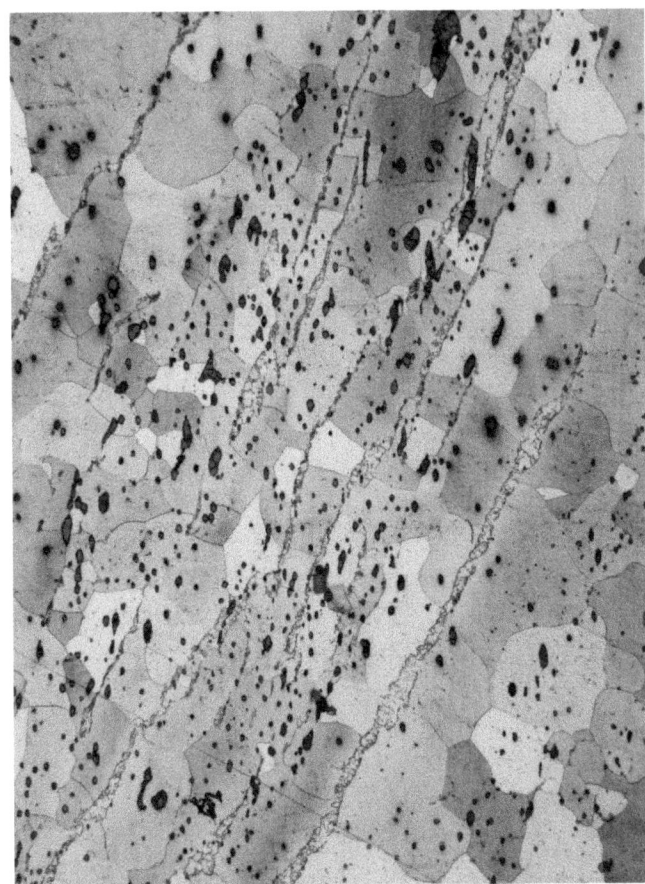

Figure 2.2.4 : Blade of a Gallic sword forged with shifted plates. Micrographs of a transverse section of the blade (3.9 cm) and magnification.

The matrix is ferritic with large grains and low carbon content, practically pure iron since there is no pearlite. Coarse, black inclusions are oxides aligned along the welding track. Fine gray lines were identified as clusters of carbides. The blown out shapes suggests that they were scratched and incorporated to the matrix during forge welding. Sample of CCSTIFM, Jarville, Fr. ; micrographs Grenoble INP, Fr.

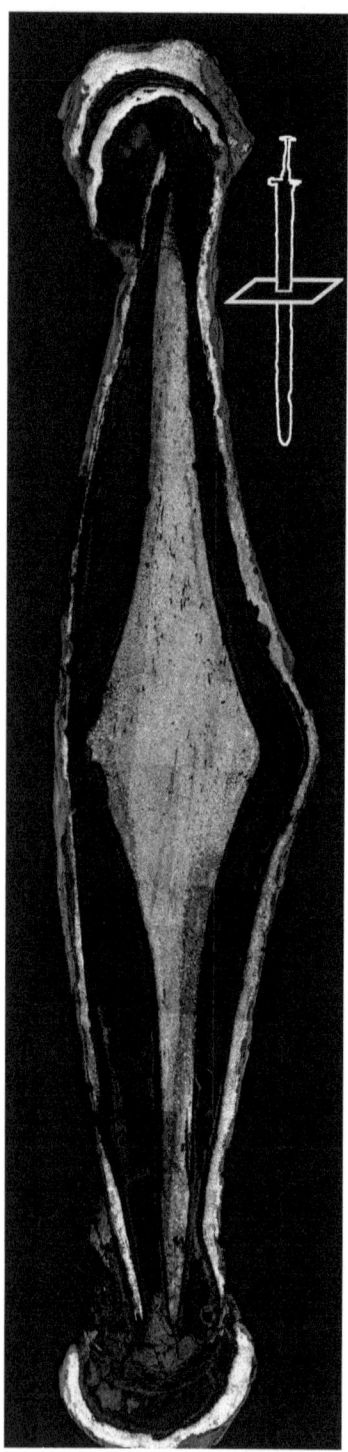
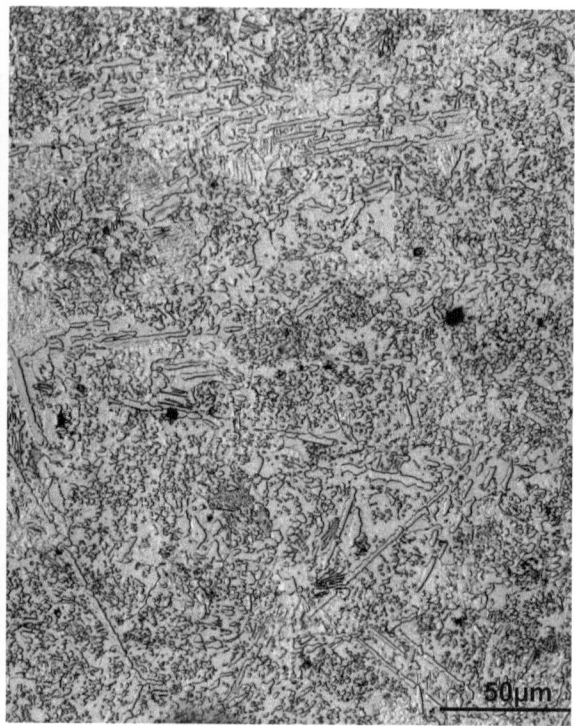

Figure 2.2.5 :
Gallic sword with sheath found in Châlons-sur-Marne (La Tene I). The section is a single part including the cutting edge, 4 cm in length and 0.4 cm in thickness in the largest part of the blade. The steel is homogeneous with some rare tracks of foldings and welds. The martensitic zones are limited to the edges of the blade.

The optical micrograph of a zone taken in the central part of the blade reveals the presence of coarser carbides inbetween the spheroidized pearlite, which lets suppose a percentage of carbon of at least 0.9. Sample from CCSTIFM, Jarville, Fr, micrographs Grenoble INP, Fr.

2.3 The Etruscan swords

At the end of the second millennium, people of Asia Minor fled the Phrygians and reached the North of Italy giving birth to the first civilization of this region. It was the Etruscan civilization which developed between the 10th and 5th centuries BC. By the 6th century, the zone of influence extended to Rome, then declined until the first century BC. This civilization appeared very original by various aspects religious, artistic, political and technical. The origin of the Etruscan people remains still mysterious and its language does not resemble any other.

The Etruscans practiced oversea trade and developed luxurious craftmanship particularly in the field of ceramics and metallurgy. The metallurgy of iron was supported by the presence of many mines of easy exploitation, in particular on the Isle of Elba opposite the coastal city of Populonia. According to the archaeologists, the reduction of the ore on site would have caused the deforestation of the island and imposed later the transfer of this activity onto the continent.

Tombs dating back to the 4th century BC contain beautiful mural paintings representing typical everyday scenes, banquets, games etc. The two characters sketched in Figure 2.3.1 show the weapons.

The sword drawn in Figure 2.3.1 is so corroded that it looks like a fork with four teeth. The four separated bars clearly reveal how it was built. Metal fragments projecting on both sides of the extreme bars let suppose the presence of welded cutting-edges.

The studies mentioned by Pleiner [39] establish that the Etruscan blacksmiths practiced the welding of steeled layers. This sword made up of bars welded side by side looks very much like the swords of the Celts. There have probably been reciprocal influences between the Etruscan and Celtic peoples either by trade, or during conquests.

Figure 2.3.1 :
Sketches based on characters on low reliefs dating from about 360 BC presented at the Paestum Museum, Italy.
The Etruscan warriors are equipped with a spear and a shield, a crested helmet decorated with feathers and with armor.
The character who holds up a sword seems to be a civilian.
See opposite a sketch by the author representing an Etruscan sword shown in the Museum

3 *The swords of the first millennium AD*

The figurative blades appear in Europe as a natural evolution of the primitive composite structures. It was at the same time as a metallurgical technique and an art that developed and culminated between the 8th to the 10th centuries.

3.1 Merovingian (481-751) and Carolingian (after 751) swords

From the west of England to the east of Russia, a single type of sword

The spear is a fighting weapon, but the sword remained the most widespread individual weapon from the 5th to the 10th centuries, between the period of migration and the medium Viking age. Swords were found in all of Europe, from Yugoslavia to Scandinavia. Shapes have evolved over the centuries. From the 8th century they have the characteristics of swords known most often as Vikings swords *i.e.* straight sword, double edged. The blade is wide making for a heavy weapon (Figure 3.1.1). The various characteristics such as length, shape of the pommel and guard are features which roughly enable defining the period and place of production. Amongst these swords, only a small proportion are figurative blades with a design built voluntarily by the alternation of different metallic layers.

Figure 3.1.1 : Warrior carrying a massive sword.
Courtesy Moebius

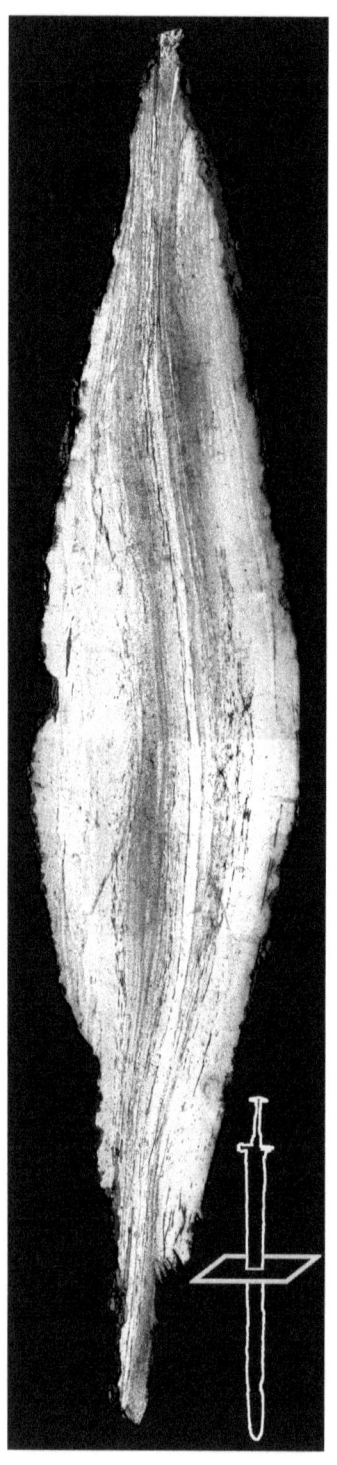

The swords found in the north of France and Germany have a wide variety of designs, classified into 17 patterns that probably correspond to different workshops of master blacksmiths and at different times [7, 32, 41-42].

The first example selected does not have any pattern, on the other hand it proves the continuity of the practice of forging by currying laminated layers as the Celts did. In Figure 3.1.2. the cross section of a Merovingian spearhead is represented. Attentive examination reveals a more evolved know-how. The welds are less coarse and the arrangement of layers is designed by combining steel grades with different carbon contents. More carburized zones appear on the cutting-edges in which a martensitic structure can be detected

Figure 3.1.2 :
Undated Merovingian spear.
Micrograph corresponds to a cross section of 3.5 cm width and 0.5cm thickness in the widest part. Sample from CCSTIFM, Jarville, Fr.
Micrograph Grenoble INP, Fr.

The second example is chosen to illustrate the major difference with the forging of the Celtic blades which is the deliberate use of steel plates of different grades to create decoration. It is the combination of a medium carbon steel and a very low carbon iron which often contains phosphorus. The effect of the presence of this element is discussed in Part II of the book.

The Merovingian sword in Figure 3.1.3 appears to have been built from the welding of a stack of shifted plates. During forging the shift generates oblique layers. The plate obtained is stretched, folded up and welded thus creating a chevron pattern visible on the section of the blade.

Figure 3.1.3 :
Undated Merovingian sword.
Sample from CCSTIFM, Jarville, Fr.
Micrograph Grenoble INP, Fr.

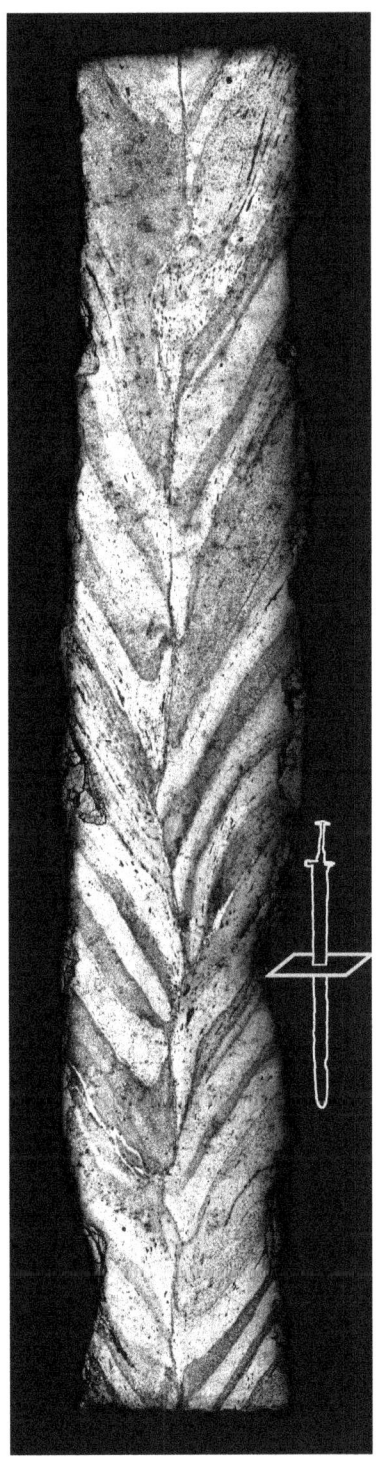

The various types of figurative blades

For simplicity, only two main types of decorated blades are presented in this book. The first type, called A here, corresponds to blades made up of bars welded side by side forming various patterns, depending on the arrangement of chevrons and twists. The pattern of each constitutive bar is visible on the two sides of the blade. The cutting-edges are welded laterally. The second type, called B, is a structure in which the decoration is a plating, a metal marquetry.

The Merovingian sword in Figure 3.1.4 is of type A. The body of the blade results from the welding side by side of two bars laminated and twisted. Two cutting-edges are welded laterally. The micrograph of the twisted zone reveals two grades of steels (see also Figure 9.1.3). The fine dark lines inside the two layers are oxide alignments proving successive foldings.

Figure 3.1.4 :
Fragment and electron micrograph of Merovingian sword (5-7th centuries) found in river Saône near Chalon.
Courtesy Musée de Chalon sur Saône, Fr.

The second example of a type A sword shown in Figure 3.1.5 is a complete classic Carolingian sword found in Strasbourg in 1899. The dating locates it between 780 and 950 [42].

The upper part shows a chevron pattern at the center of the blade, obtained by welding two bars twisted in opposite directions.

The lower part shows a series of waves parallel to the axis. The length of the sword is 94.5 cm.

The welded sharp edges never have a decoration.

Figure 3.1.5 :
Carolingian sword.
Courtesy Musée Archéologique de Strasbourg, Fr

The third example of a type A sword (Figure 3.1.6) is a Merovingian sword found in a necropolis of the Jura. Due to deep corrosion the structure of this blade is not easy to decrypt. In the Museum, a video is presented together with the sword, showing Pierre Reverdy manufacturing a similar blade. The steps are summarized in Figure 3.1.6. The smith arranges a series of seven flat sheets, ones in soft iron, the others in carburized iron, then he stacks them alternating the grades and welds the unit forming a bar. Then the bar is forge stretched and hammered so as to obtain a smaller laminated bar with an almost square section. The bars are then hot worked, folded up like an accordion or twisted with a more or less tightened turn of screw. Two or three of these rods are forge welded together. The distribution of the various metallic layers lets a pattern appear after polishing and etching with a concentrated solution of ferric chloride.

The swords of the first millennium AD

Figure 3.1.6 :
From left to righ: - scramasaxe of the Merovingian necropolis of Crotenay (Jura) dated 580-600. The decoration is hardly visible on account of deep oxidation, in spite of two restorations;
- enlarged zone of the blade; Courtesy Musée Archéologique de Champagnole, Fr.
- steps of the manufacturing of the blade by P. Reverdy, Romans, Fr.

The blades decorated by plates, a kind of marquetry

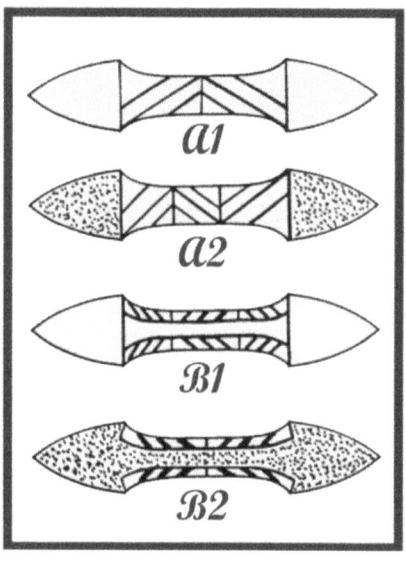

Figure 3.1.7 :
Schematic cross section of various types of blades. Separate steel edges are welded in A1, A2 and B1.

The center of B type swords consists of two composite steel plates which are welded together or onto a soft iron core. Both faces of the blade can be decorated by different plates. A sketch (Figure 3.1.7) shows four assemblies among others. The presence or not of an iron core does not seem to be related to one period, but rather to a workshop.

There is no available transverse section for the examples of Figures 3.1.8 and 3.1.9. In certain cases, the careful examination enables proving that a decorated plate had been welded either by the presence of bare parts, or simply because patterns on both sides of the blade are not connectable. X-ray examination reveals the presence of layers (Figure 1.4.3 in which the D1 image appears to be of type A whereas D2 more blurred may be of B type).

3.2 The Vikings swords

The problematic origin of the swords of the Vikings

At the end of the 8th century, thousands of the men living in Scandinavia embarked on ships to conquer and plunder new territories, they were the Vikings. They crossed the seas and penetrated all of Europe by river with their famous ships adapted to any navigation [43]. Vestiges attributed to the Vikings are numerous in Scandinavia and in the British islands. They are rarer in France. Some beautiful swords were

THE SWORDS OF THE FIRST MILLENNIUM AD

Figure 3.1.8 :
Merovingian sword. The two magnified areas were not taken out of the same side of the blade, neither directly opposite one another, being chosen where corrosion was least and pattern most clearly visible. Courtesy Musée de l'histoire du fer, CCSTIFM, Jarville, Fr.

Figure 3.1.9 :
Spear head and Merovingian sword chevron patterned.
Courtesy Musée de Chalon sur Saône, Fr.

found in the Loire, Somme and Seine rivers. However, as these vestiges were lost during bloody battles, doubt about their origin remains.

The Vikings entered legend like ferocious warriors armed with swords. However, a remark in the book of Byam *Weapons and Armors* [46] corrects this popular image a little: *the ordinary warrior used mostly a long pike (lance) whereas the professional fighters and the clan chiefs usually fought with battle-axes.*

The first studies [7] concluded that the majority of the Vikings' swords came from smuggling and plunderings because there is not in Scandinavia any trace of true important production centers like those of the area of the Rhine. More recent studies seem to prove that some vestiges can result from a local production [44]. Carefully, the archaeologists mention *swords of the Viking type*, or *Viking age* [45] as those of Figure 3-2-1 which presents a set of weapons dated from the 10th to 13th centuries.

The typology of these swords has been investigated extensively in relation to the find-locations or with the period, resulting in a chronological classification into many styles according to the characteristic features of hilts, pommels and blades. A small proportion of these swords were of a figurative type (with decoration) and the silver inlaid blade of Figure 3-2-1 is exceptional. Many blades wore inscriptions such as *Ingelfred, Ulfberht* assumed to be the manufacturer brand, the name suggesting Frank's workshop. Whether these letters are pattern-welded or inlaid is still a matter of discussion. Variants of these inscriptions, maybe imitations, appear over a period of two centuries.

The microstructure of the blades has also been examined, in order to reveal how they have been made. Many changes in the steel production have occurred during this period. A small proportion of the blades belong to the pattern-welded type in agreement with the Carolingian style. In the course of the 10th century, Damascening was progressively abandoned. Meanwhile, the quality of wrought iron improved and it was available in larger blooms. The blades which were previously created using small pieces welded together, in particular an iron core and hardened steel edges, could be forged from a single piece of steel. A possibility of importation of crucible steel is discussed §11.3.

Figure 3-2-1:
Four Viking type swords and magnification of the left-hand sword decorated with silver inlaids. Courtesy Musée de l'armée, Paris, Fr.

3.3 Swords in China

Swords dating from the two first centuries BC were discovered on several sites. Rubin and Ko [19] describe a blade on which gold inserts mention the manufacturing date *under favorable auspices* and specify a number of 30, 50 or 100 refinements. Metallographic examination revealed small inclusion particles forming alternating layers distributed in numbers close to 32, 64 and 128. There is an obvious correspondence with the number of operations during the folding and forging process which leads to a number of layers that is to the power of 2. In this case, the operations were probably carried out with the aim of refinement of the metal, to improve it, and not to obtain a possible patterned structure.

3.4 Swords forged with crucible steel

The sword blade of Figure 3.4.1 is undoubtedly one of the oldest blades observed. It is associated with the tomb of Alains at Klin Yar in the north of the Caucasus and is dated from the 3rd-4th centuries of our era. Taking into account the presence of a high proportion of cementite carbides, it seems evident that the blade was forged out of crucible cast steel. The carbides are aligned, but they do not appear distributed in strips. The handle riveted with the blade presents a microstructure corresponding to welded layers. The blacksmiths of this period used smelted iron and crucible cast steel simultaneously. They probably kept the crucible steel for parts requiring the best mechanical properties.

The acquisition of a decorative texture by a strictly controlled forging and its improvement with the famous moire pattern shown § 6.6 came much later. The oldest textured blades, visible in museums go back to the 11th century according to a private communication from Dr Klaÿ, coauthor of a book about the Moser's collection [47]. Also, wootz patterns are described in Persian manuals written around the 11th-13th centuries [48].

Figure 3.4.1 : Sword of the 3rd century found at Kislovodsk, Russia and micrographs of the blade made out of crucible steel
Top : optical micrograph after etching, the abundance of cementite precipitates reveals a hyper eutectoid composition ;
Bottom : electron micrograph showing carbides in clear contrast.
Courtesy A. Feuerbach [12].

4 Did you say Damascus steel, Damascene or damask ?

4.1 The debate on the origin of the naming

Of course, the origin most often proposed for Damascus steel comes from the name of the town in Syria and dates back to the beginning of the second millennium at the time of the crusades (1095-1270). The crusaders discovered culture, practices and objects not found in their countries, and in particular two-edged swords of their enemies which were sometimes returned as spoils. The Eastern swords acquired a reputation of lightness and strength, with incomparable blade decoration with a one million wavy lines and a famous cutting edge. The idea that the swords were so different from the heavy Franquish swords was essentially spread by Saladin legend, a legend revived a long time after the crusades in a novel entitled *Talisman* written by Walter Scott, a romanticized history which became very popular. Specialists of weapons contest this approach, they mention that most of the swords used by Muslims during the time of Crusades were in fact straight and double edged [49].

Symmetrically, the Arabic commentators gave the name of *swords of Köln* to the Frankish two-edged weapons which were also very appreciated in Muslim countries as spoils of war [7]. Damascus was, as Köln, a center for exchanges. Eastern weapons, forged in Damascus or somewhere else, were effectively exported from Damascus in the period from 900 to as late as 1750.

Other assumptions for the naming *damask* propose a deformation of an Arabic word meaning water because of the wavy, swirling pattern. Suggesting the same image, the term *ondanique* was found in certain narratives. Another proposition is that the name seems to be the distortion of the Greek *adamas*, or of the Persian *aimas* justified because the steel was considered as hard as diamond.

When and how the word *damask* became a reference to a certain type of steel remains mysterious. However it is clear that the use of this name in Occident is largely posterior to the time of the crusades. However, all of the dictionaries refer more or less explicitly to steel from Damascus or at least forged in Damascus as it appears in Table 4.1.1 reporting some definitions of the word *damask* together with the French word *Damas*, equivalent to Damascus.

4.2 From laminated to figurative blades

In chapter 3 it is explained how blacksmiths forge welded several strips to build a blade. They probably realized that alternance of different grades could improve the mechanical properties possibly as early as the middle of the first millennium BC in the case of Etruscan smiths. The great innovation of the first millennium AD was the use of welded layers of different grades to obtain a decorative aspect. How did one name such decorated blades in the Middle Ages? Certainly not damask like now, probably *figured* or *figurative* because the expression is in ancient dictionaries with the meaning *which has a pattern*.

The words *damassé* or *damas* and an alternative *dapmas* appear several times in French texts in the 14th century around 1352 used always to describe fabrics. All French or French-Latin dictionaries give as the first meaning the decorated silk fabrics woven in Damascus. The particular weaving of such silk fabrics was probably developed in China but it was practiced in Damascus where it gained the reputation. By extension, this name now indicates fabrics of silk, wool, linen or cotton woven with monochromic threads and whose weaving forms with a contrasted reversible pattern.

After the crusades and until the 16th century there is an obscure period about forge welded steel. A simplifying assumption is that the figurative steel became old-fashioned in the West and completely given up. The manufacturing of decorated swords probably decreased for economic reasons because the last crusades had been ruinous and they were followed by disturbed historical periods. Moreover, the production of steel became industrialized starting from the 14th century and technical evolution made it possible to manufacture strong blades with finer homogeneous steel. In addition, the use of weapons evolved with the

Table 4.1.1 : Some definitions for damask and Damas (Damascus in French), with original spelling, and available for consultation on the Internet.

1636	Dictionnaire français-latin de Philippe Monnet	**Damas**, étofe de soie, invantée à Damas, toute **figurée** **Damassé**, **figuré** à guise de Damas
1762	Damas Dictionnaire de L'Académie française 4me édition	*Damas, étofe de soie à fleurs...* *On appelle un acier de Damas Un acier d'un trempe excellente,& dont il est fait un grand débit à Damas.* *Un sabre d'acier de Damas...* Damas, flowered silk fabrics We name "Damas" steel an excellent quench hardened steel, which they use a lot in Damascus. A "Damas" steel saber...
1828	Damask Webster's Dictionary	A silk stuff... Damask-steel, is a fine steel from the Levant, chiefly from Damascus, used for sword and cutlas blades.
1832	Damas Dictionnaire de L'Académie française 6me édition	***Damas, (on ne prononce point l'S)*** *étoffe...* *Damas se dit encore d'Une lame faite de cet espèce d'acier très-fin et renommé par l'excellence de sa trempe, qui se fabrique à Damas ou selon les procédés employés à Damas. Ce sabre est un damas, un fin damas, un vrai damas. On dit de même, Acier de Damas.* Damas (say dama) fabrics... Damas is said also about a blade of this kind of highly refined steel, famous for its quench hardening, which is manufactured in Damascus or according to the processes used in Damascus. This saber is a fine Damas, a genuine Damas. We also say Damascus steel
1913	Damask Webster	Damask steel, or Damascus steel, steel of the kind originally made at Damascus, famous for its hardness, and its beautiful texture, ornamented with waving lines; especially, that which is inlaid with damaskeening; - formerly much valued for sword blades, from its great flexibility and tenacity.

practice of fights and, in particular, with the common confrontations in duels.

Maniere de faire l'Acier façon de Damas.

Nous aurions pu parler de cet Acier au Chapitre où il a été queſtion des étoffes, parce que c'en eſt une; mais comme elle n'eſt pas propre à faire des tranchants fixes, nous avons cru en devoir traiter à part, quoique cet acier ſoit très-bon pour faire des Couteaux de table.

Cette étoffe qui imite le Damas naturel, à s'y tromper, devient coûteuſe, tant par le temps que par le charbon qu'il faut employer, & par la diminution de la matiere; car ſi l'on veut en faire 3 livres peſant, il faut prendre 6 livres de matiere.

Figure 4.2.1 :
Top: Extract of Perret's text entitled "Manner of making steel in the damask way" which assimilates steel to a material.
Bottom: Smith according to a detail of an illustration [50].

Blacksmiths and academics do not speak the same language!

The western smiths asserted the names of *Damas* and *damassé* for their welded blades made in the tradition of the Merovingian period and claimed to forge the authentic damask. Perret in 1771 [50] also proves by his text in Figure 4.2.1 that he knew that the oriental blades were made starting from crucible melt steel *Damas naturel*. But he also asserted that the smiths successfully made by currying, blades just as beautiful and even better than the oriental swords, because they were less brittle. There was assimilation of the vocabulary used for the fabrics. Moreover the words materials and fabrics *(étoffes)* appeared about steels in the texts of Perret and of Landrin [52]. The analogy is reinforced by the possibility of obtaining an effect of relief for steel as for fabrics. The use of words from other fields is common, for example *currying* adapted primitively to leather working and used later for iron.

Clouet in 1803 [51] and Landrin in 1835 [52], described the process of manufacturing the *damas*. In the report on the manufacturing of the *lames figurées dites damassées*[1], Clouet defined how to obtain the multiple decorations including decorations in waves, undulations and the watered effect. Everything is explained : the way of constituting a stack, the choice of grades, the tricks to create patterns, the heat treatments, the chemical etchants, the pastes for polishing, the coatings of protection, the use of borax and even the incorporation of rare metals such as platinium. Let us point out that the described methods are those still employed by contemporary blacksmiths.

4.3 Crystallization steel or pattern-welded steel

The Egyptian Campaign (1798-1801)

The Turkish armies besieged Vienna in Austria in 1683, giving the opportunity to European warriors to worry about oriental weapons. But it is the campaign of Egypt headed by Napoleon Bonaparte, that marked a second cultural confrontation and aroused in France a strong interest for all that came from the Middle East. The archaeologists who visited Egypt prolonged this influence. Thus, Champollion received in gift an oriental sword at the time of his stay in Egypt. De Luynes writes in a report published in Paris in 1844 on the manufacturing of the damasked steel more than forty years later : *After the expedition of the French people in Egypt, the damask steel of India and Persia had acquired a so great reputation, as we tried hard to imitate it, and to make bladed weapons which quality and the brambly vein could compete with those which characterize the oriental weapons.* Clouet tempted it, but no currying of this nature imitates exactly the damasked aspect.

Yet, the laminated structures imitated sometimes so well the design of the wootz that it was difficult even for experts to make the distinction between the welded structures and the structures obtained by crystallization. Moreover, numerous blacksmiths believed in good faith to have found the forging of the oriental swords.

---

1. figurative blades known as damasked, to be remarked the explicite mention of the word "figurées"

The search for resisting steels : 1800-1850

Around 1750, the reputation of Eastern swords weakened, it was said that the technique was lost, the blacksmiths had forgotten the process. This historical aspect is controversial and some people think it is a legend. It was also argued that the specific ores to manufacture wootz were worked out in India. However it is evident that forging high carbon steels is difficult (chapter 10) and can be done successfully only by well experienced smiths. Then facing a high demand possibly resulted in poor quality production.

In England, the study of Indian steel and Eastern damask structure began in a scientific way in 1795 with Pearson seeking an objective understanding of why oriental steel was so successful. Michael Faraday carried out with James Stodart around 1818 a series of experiments with addition of elements to steel in order to make them more resistant to corrosion, but as they had tested some rare and expensive metals they were not able to end to a larger-scale development.

In France Jean Robert Bréant published in 1823–24 the first descriptions of the microstructure of oriental damasked steel and confirmed that its essential characteristic is a high carbon content. The explanation was reported by Landrin : *a steel in which, by the effect of a suitably arranged cooling, it took place a crystallization distinct for iron and carbon.* A study published in 1824 shows that MM Stodart and Faraday succeeded in obtaining by fusion a steel which once forged *became agreeably damasked.* An important conclusion is *that the figures which one notices on the Damas are thus a phenomenon of crystallization.* Because *the wootz does not lose this property by a new fusion.*

The success of the experiment was attributed only, and wrongly, to the use of silica and alumina for the fusion. Undoubtedly, these oxides played a role by the formation of a slag protecting the melted metal at the bottom of the crucible. In Russia Pavel Anossoff devoted his life trying to find the secret of this manufacturing. He sought to determine best clays and graphites and, like Bréant a few years before, he tested all kinds of additions to iron, including diamond!

From the beginning of the 19th century, the blacksmiths knew how to distinguish crucible cast steel, which they called crystallization steel, from welded damask. The ambiguity lasted until the 20th century, probably maintained because the Eastern blades were extremely well appreciated. However let us not retain a simplified image in which all that was manufactured in the East was made out of crucible steel. Construction by welding was also practiced and in particular the welded damask made it possible to manufacture rifle barrels in the 17th century with the technique of rolling up which was to be imported and developed in Europe (§ 6.4), in particular in England, Belgium and France under the label of *damas* in the two later countries. This name is still used in French for figured steel obtained by currying and presently tends to refer in an excessive way to indicate laminated structure. Since the 20th century the names *Schweissdamast* in German and *pattern welded* in English are used and their meaning is quite clear [16].

4.4 Damaskeen, damaskeened, damascene

The craft art consisting in inlaying silver or gold decoration is very ancient since it was practiced before the iron age among the Egyptians, the Greeks, the Romans, the Japanese and the Chinese. Some examples are presented in Figures 3-2-1, 4.4.1, 6.2.1 and 6.6.2. The Arabs developed this technique to decorate and write inscriptions on sword blades made in Damascus. The primitive technique consists in carving a groove using a chisel. Once a drawing was carved, then a golden or silver thread was inserted by means of a punch with a flat base striking with a small and light hammer with a wide mouth.

Carving was much more difficult to carry out with Damascus steel than with other metals because this steel is hard, wear resistant and less ductile than precious metals. The Arabs brought this know-how with them to Spain where it is still practiced for manufacturing mainly jewelry or decorative objects. Toledo is the world's largest center of production.

All the words refering to this craft as damaskeen or damaskeened and similar in other languages such as *damasquinado* in Spanish, *damasquiné* in French, and *damas-ziert* or *Damaszener-Arbeit* in German seem to focus upon the name of the town of Damascus. The author al-Beruni mentions swords made by a smith he names Damashqi, or signing this name which means from Damascus. In a thesaurus English

Figure 4.4.1 :
Damaskeened cutlery with silver and gold inlays. Private collection Stora

dictionary the example chosen for the Damascus steel is a sentence extracted from *El ingenioso caballero don Quijotte de la Mancha* of Cervantes published in 1615: - *en lugar de espadas truxessen* cuchillos tajantes de **damasquino azero***...* (* old spelling); translated by John Ormsby - *in place of swords wield trenchant blades of **Damascus steel**...* Here the word damasquino is taken in the broad sense of Damascus steel.

A study into the Spanish names ending in "ín", "án" and "í" demonstrates that many names of fabrics, ornaments and clothes have a toponymical origin [53]. Then *damasquino* and derivative words mean pertaining to, or originating at, the city of Damascus. Damascene has also the meaning "who/which comes from", for example St John Damascene.

5 *From swords to knives*

Numerous objects of the domestic life were manufactured out of wrought iron. Can we compare the manufacturing of the common knives to that of the blades of swords or to that of spearheads? The excavation of a site of the 11th century testifies to the know-how of the blacksmith-knifemakers in a rural environment.

5.1 The lake of Paladru

A site preserved by its immersion

The vestiges of a fortified habitat of the year thousand were discovered in 1972 in Charavines, in Lake Paladru in France (between Grenoble and Lyon). Since then, the site has been methodically explored. The sub-aqueous character of the deposit facilitated a remarkable preservation of the organic rests and the light basicity of waters protected the metal objects from oxidation.

A group of emigrants in about the year one thousand

During the second half of the 11th century the climate became a little drier and, undoubtedly, slightly warmer. The living conditions improve because of the end of the invasions and, also, the climatic improvement supported better harvests, therefore a demographic growth occurred. Consequently, the conquest of new lands to be exploited became necessary to feed the increasing population. At the same time, the old social structures inherited from the Carolingian period were gradually replaced by the feudal organization of society in which the authority was scattered and split into multiple autonomous cells.

Around 1006-1009, three families, probably encouraged by the archbishop of Vienne (south of Lyon), arrived in the area to establish new soils. "Families" must be understood in the broad sense of the word *i.e.* including their domesticity. These colonists found on the border of the lake an accessible chalk beach, recently emerged with the

Figure 5.1.1 :
The equipment of the knight presented at the Musée de Paladru, Fr

decline of the water. They chose this place to build their environment and cleared the surroundings. They developed farming, breeding, fishing and hunting. They lived a widely autonomous life by practicing all the craft: work of wood, leather, wool and metallurgy. The forest supplied an immense wooden reserve to create forges. The housing settlement looked like a vast farm distributed in several buildings. The men and the animals protected by a fortification made up of a palisade with a covered way. Since 1020 however, the rainfall increased and made, little by little, the lake level rise. Between 1038-1040, the men had to abandon the site and move towards other zones, undoubtedly close because the surrounding soil remained exploited.

Knight-farmers

The archaeologists Michel Colardelle and Eric Verdel [54-55], collected equipment of knights and weapons which form the panoply of the horseman in weapons of that time (Figure 5.1.1).

They estimate that these colonists living at Colletière were knight-farmers, *villani caballarii*. These colonizers of a region without border had to be able at the same time to exploit their territory and to defend it militarily if necessary. They were thus equipped to ride a horse and to participate in war raids.

5.2 The knives at the beginning of the millennium, in the year one thousand

Blacksmith-knifemakers ?

Approximately four hundreds blades of knives are found among the vestiges. This amount is much too large for only local consumption, it lets suppose a commercial oriented production. On the other hand, only one pommel of a sword, some pieces of spearheads or javelins and the waste of a forging mill were discovered on the site. Swords were found close to Paladru, but are not a part of the exactly dated vestiges located in the lake. Did the smiths make or did they just repair the weapons? Were the weapons taken during the move because they were more precious than knives? It is possible because a sketch of a spear was recently discovered on the site.

In the 11th century, after Merovingian and Carolingian periods, the process of forging by the welding of several steel plates was well-known. However, as they produced knives, common objects, it was not necessary to perform decorations similar to those of the sword blades. In fact, the largest number of the blades seem to be of the laminated type with a reported edge, as already observed on the Gallic swords § 2.2.

What reveals the microstructure of the blades

From the surface aspect, blades have a laminated structure (Figure 5.2.1). The sharp edge appears darker all along the lower part of the blade. The images of Figure 5.2.2 present the corresponding cross sections. The micrographs reveal stacks of layers, longitudinally welded to improve the resistance of the blade to a transverse rupture. Each layer is characterized by a certain grey contrast in relationship with the proportion of pearlite linked to the carbon content. In a magnified zone, two layers are limited by fine continuous lines of oxides. A characteristic banding inside each layer is observable, it occurs after a long time forging with particular thermal conditions (a point to be discussed in details in §11.1)

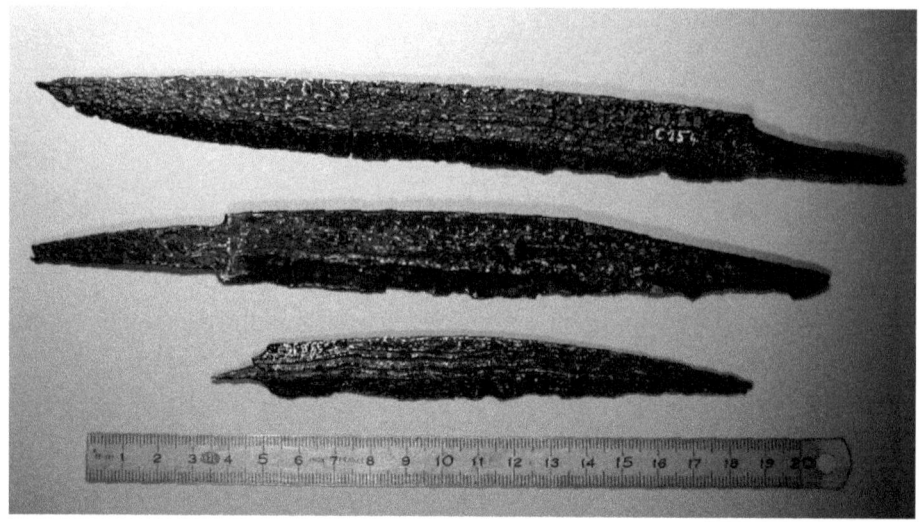

Figure 5.2.1 :
Blades of knives found at Paladru, the dark parts at the bottom of blades are perhaps the reported sharp edges.
Collection Musée Dauphinois, Grenoble, Fr.

The steel plate, corresponding to a layer, is obtained after many processes of *folding then forging*. During these operations the metal surface is submitted to hot oxidation and has to be protected by a flux. For that purpose, mixtures easily fusible, made of sand, of clay or of powdered siliceous scorias are employed at that time to form a protective coating of oxides.

A part of these oxides remains trapped between the layers as the strings of inclusions shown in Figure 5.2.3. The oblique shape of the edge facilitates the expulsion of slags from the weld. Certain tracks of welding appear as very fine, continuous and regular lines of oxides while the others are marked by a succession of small globules. This difference is due to the fact that freshly formed lines are continuous then, during the multiple heatings, the former lines split up in short plates, or even globules which spheroidize.

Some blades were observed in the Laboratoire d'Archéologie des Métaux de Jarville [54-55]. The diagram of Figure 5.2.4 summarizes the various models found for the welded assemblies. The most

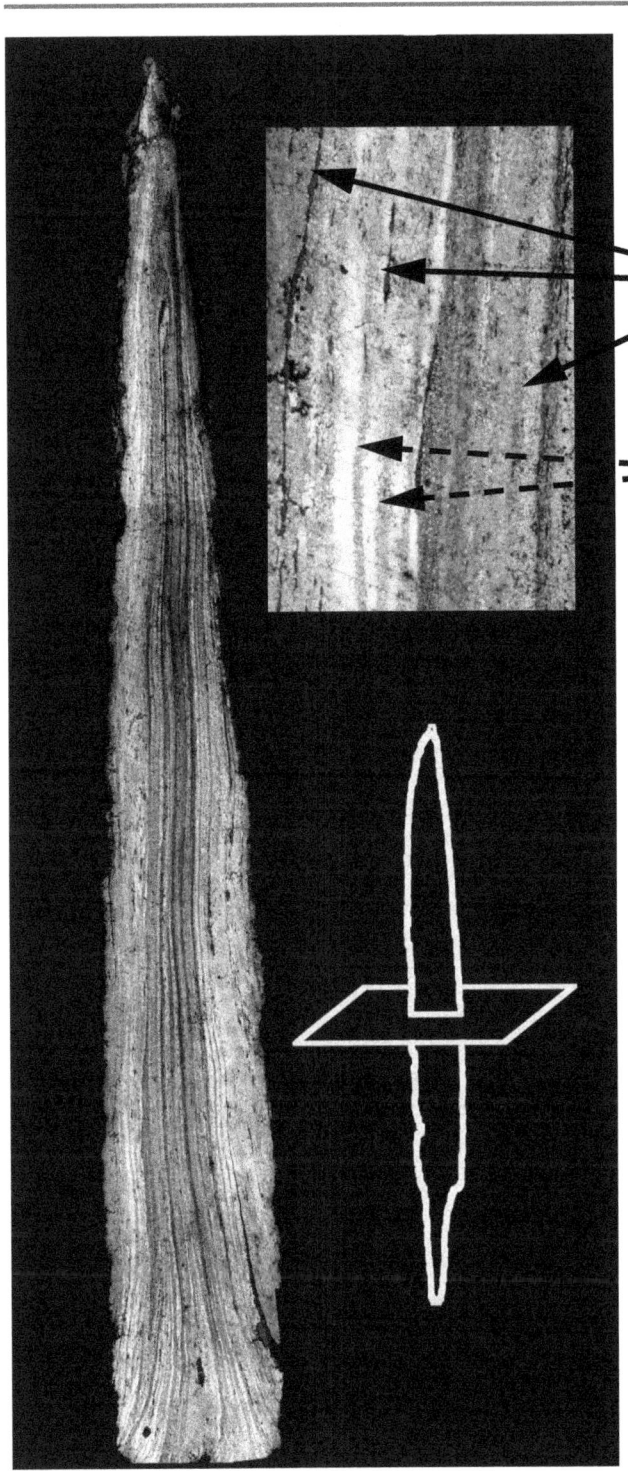

weld tracks

banding

Figure 5.2.2 :
Optical micrographs examined in a plane perpendicular to the blade and magnification of an area. Samples of CCSTIFM, Jarville, Fr, Micrographs Grenoble INP, Fr

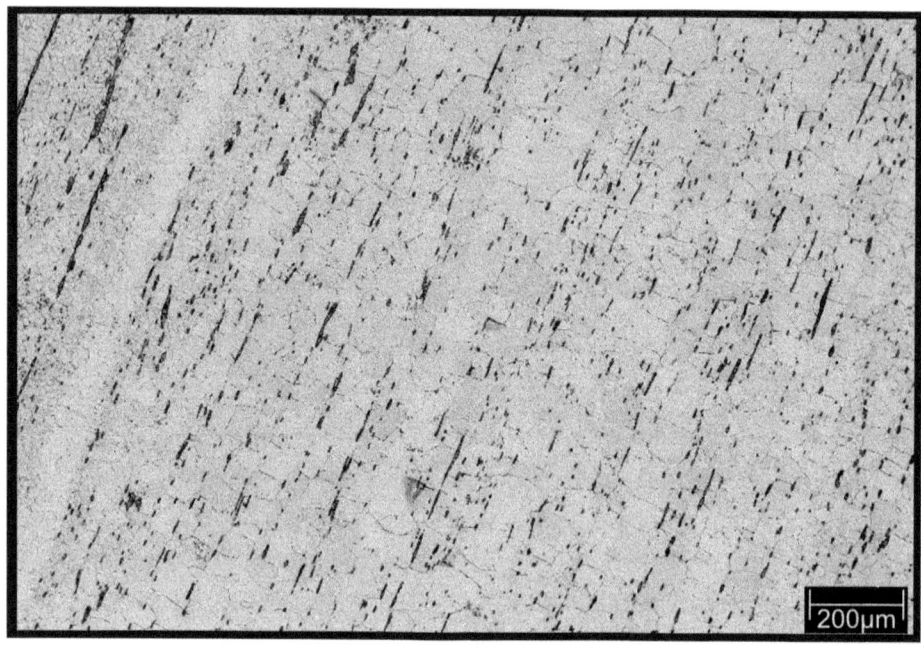

Figure 5.2.3 :
Electronic image of a blade which reveals oxide inclusions entrapped between the layers.
The great number of oxide alignments shows that this blade was particularly worked for a long time with many foldings. Other blades have coarser inclusions. A clear band can be seen on the left side of the micrograph, this is ferrite formed on both sides of a weld in contact with silicates resulting in a decarburizing effect.
This is a backscattered electron image which highlights a contrast of density, the oxides which are less dense than the matrix appear much darker while the carbides which are approximately as dense as the matrix are hardly detectable.
Courtesy Grenoble INP, Fr.

common model is the C type with a laminated body for the blade and a reported edge. The models E and F seem reserved for shorter blades. The model B evokes the pattern of a damask sword in reduction (see Figure 3.1.3). Was there research for damask decoration? Possibly! But, more simply, it may result from a folding to recover a blade of type A thinned too much. The models D1 and D2 let appear on the surface of the blade a laminated aspect with wide strips.

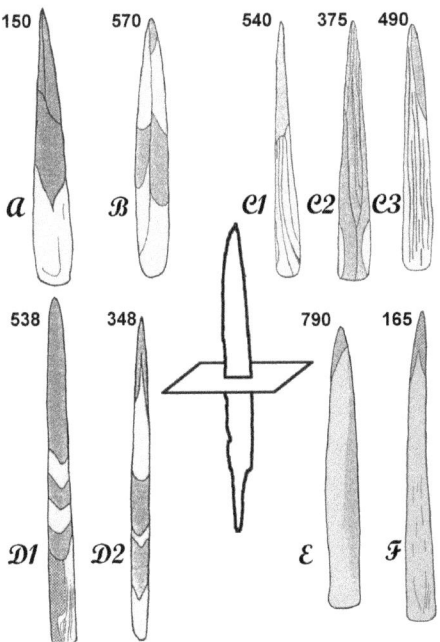

Figure 5.2.4 :
Sections of blades observed in a plane perpendicular to the blade as illustrated in the central diagram. The sketches respect the relative scale of the widths of the samples. The darker zones correspond to layers richer in carbon. The figures are measurements of Vickers microhardness Hv on the edges. Blades A, B and C are the longest with a length of about 150 mm excluding the length of tang. The maximum widths lie between 15 and approximately 20 mm and the thicknesses between 2 and 3 mm.
Document adapted after C Forrières, P. Merluzzo and A. Ploquin [55].

Unfortunately the conclusive examination involves partial destruction of the blade.

The knives were found deprived of the handles except those keeping wooden tracks. It is possible that a horn was used for the handle and has been destroyed by the course of time by hydrolysis of collagen.

The shapes of the blades are functional, completely similar to those of contemporary knives. However such knives with a tang shoved into a hole of the handle are not rigid enough and end up with the blade taking play and releasing. Then, the blade must be equipped with a new handle. Thus, an assumption is conceivable, it is that the workshop of a forging mill ensured the maintenance and the repair of the knives. Certain different original parts, sometimes broken, were on standby for repair or recovery of iron at the time of the great departure and would have been abandoned. Could this assumption explain why there is such a great diversity of form, size and especially of modes of forging?

Blacksmiths who controlled the technique of hardening

The majority of the blades have edges hardened up to hardnesses ranging between 400 and 600 Hv, except for a few blades whose edge is completely softened lower than 200 Hv. The microstructure is characteristic of a more or less tempered martensite in all the carbon-rich zones, either the edge or the body of the blade. These observations show a practice of a hardening step, deliberately followed by a stress relieving treatment. Probably, the ancient blacksmiths had empirically understood the interest to let the part cool near the hearth of the forging mill. Hardening by quenching is delicate for blades with reported edges because of the risks of rupture along the weld zone. Blacksmiths able to succeed were undoubtedly experienced craftsmen.

6 *The swords of the second millennium*

The smiths acquired the technique of forging the steel, then they were able to free all their creativity. In every country they adapted and devel- oped their traditional style by taking into account new models brought by trade or invasions. This chapter proposes a world tour of the typical creations classified according to the technique used: either forging and welding, or forging of crucible steel. However this classification does not coincide with a defined geographical limit because both techniques were practiced in certain regions, in particular in the Middle East. Lastly, the Japanese swords deserve special attention because they are the outcome of an original and very sophisticated know-how.

6.1 The use of the sword in Europe

In the time of the crusades

From the year one thousand the smiths are able to forge beautiful swords (generic term) as well in Orient as in Occident. The shape of cutting weapons evolved throughout this millennium according to their destination, *i.e.* types of fights. The great crusades at the origin of an expansion in armament, the bearing of the sword was not any more reserved for feudal lords. It is the weapon of the crusader. Also the protection of the warrior improved, the coat of mail was replaced by a more resistant armour with articulated plates. At the beginning of the second millennium swords were used essentially as a stabbing weapon, and were held with only one hand. To pierce this armour more massive swords or claymores were preferred from the 14th century. They could be handled only with two hands and by robust soldiers (Figure 6.1.1). The blade could reach 1.60 m and weigh 7 kg, the record is 2m [46]. The point of these blades became sharper to penetrate at the joint of the

Figure 6.1.1 :
Fights with two hands swords.
Sketch according to an old engraving.

plates of armour of enemies. These heavy swords were still used in the tournaments and the duels in the 15th and 16th centuries.

From the Renaissance

The sword gradually lost its main role to become an attribute, or an emblem of rank or social status of soldiers, notables and rich middle-class persons. Besides, some are richly decorated swords of ceremony.

Parallel to the expansion of the bearing of the sword, the skillful handling of the sword evolved into an art, the technique of the fight matured into sport, namely fencing. The practice of duels became common. Numerous gentlemen and aristocrats were challenged to a duel to wash the slightest insult which they considered as a dishonor. Several thousand members of the French aristocracy lost their life in two decades during the second half of the 16th century. The famous duel of minions (favorites) took place in 1578 in Paris. *The witnesses will get involved in the fight at the conclusion of which three men, among whom Caylus, the favorite of Henri III were fatally injured.* [46].

The duel was practiced with a weapon equipped with a fine and light blade. Another short sword with a double-edged blade was held in left hand but the main fight was led with the longer weapon or *rapière* which was used to ward off the opposite blows (Figure 6.1.2). The English term of rapier indicated already the heavy swords of civilians at the beginning of the 16th century. The name crossed the English Channel later so that the name of rapière is more commonly associated with the light swords of the 17th century.

By 1650, rapiers were replaced by shorter and lighter swords that gentlemen bore till the end of 18th century. They are essentially ornamental swords. Moreover, certain porcelain handles painted with a decoration of a gallant scene evoke more the court than a battlefield. Very beautiful objects were made in particular in Toledo, Milan and Solingen. At the end of the 18th century, the most luxurious swords were not made any more out of damask steel, industrial steel was used making it possible to manufacture swords of excellent quality. The craftsmen created a decoration by damascening, or engraving supplemented by a gilding with gold.

Figure 6.1.2 : Fighting in a duel with two swords.
Sketch according to an engraving from the 19th century.

6.2 Damask rapiers

The work of steel by currying and welding reached an admirable degree of perfection to the 17th century to produce fencing swords. The damask swords were very famous for their quality of flexibility and strength. The blacksmiths were able to achieve true artistic and technical exploits. The images which follow present entirely damask swords. The rapière of Figure 6.2.2 or tazza with straight quillons completely damasked, has a remarkable guard. In both cases the decorations of the handle, of quillons, and of the guard are in perfect harmony with the design of the blade.

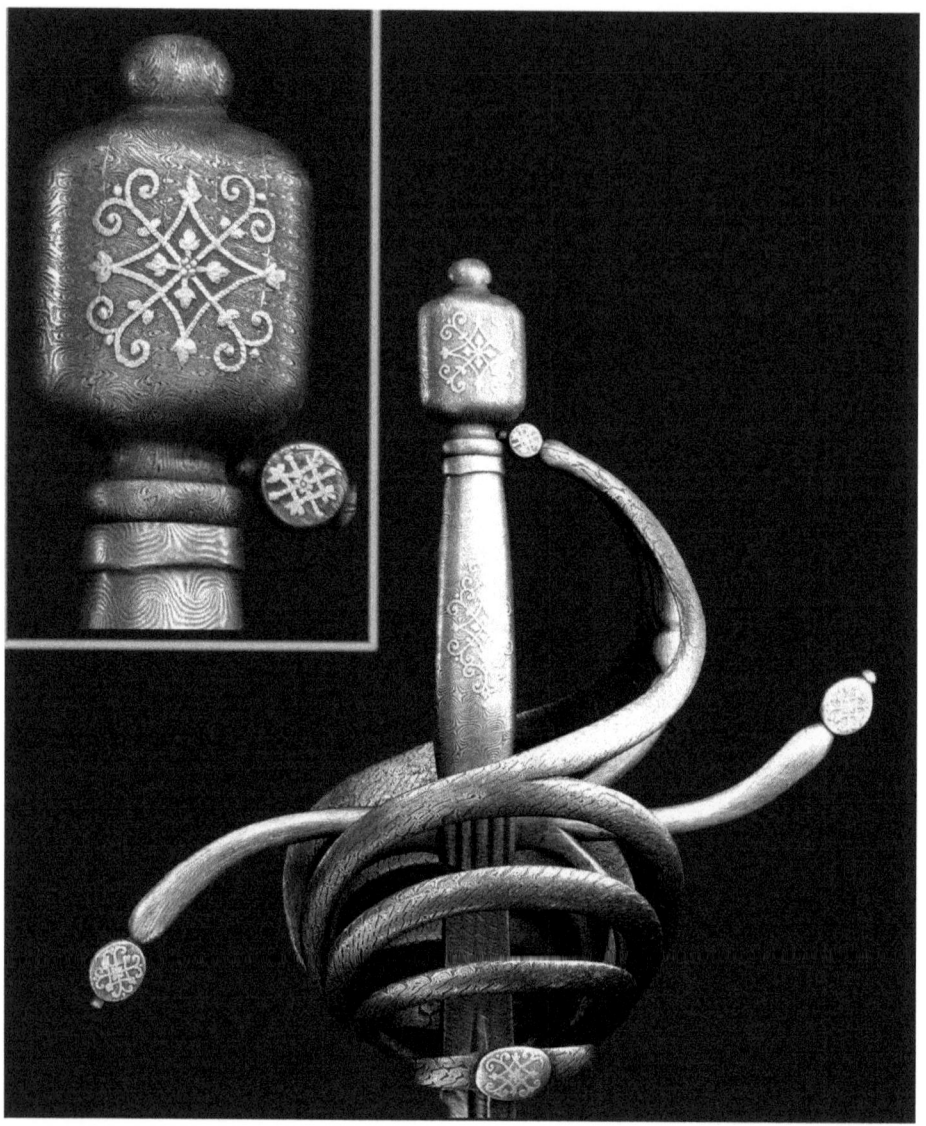

Figure 6.2.1 :
Rapier with multiple branches and two bent quillons, made in Europe in the 17th century.
All the parts of the weapon are damasked and damaskeened *i.e.* decorated with gold inlays.
(Total length 122 cm)
Collection Musée d'Art et d'Industrie, Saint Etienne, Fr.

THE SWORDS OF THE SECOND MILLENNIUM

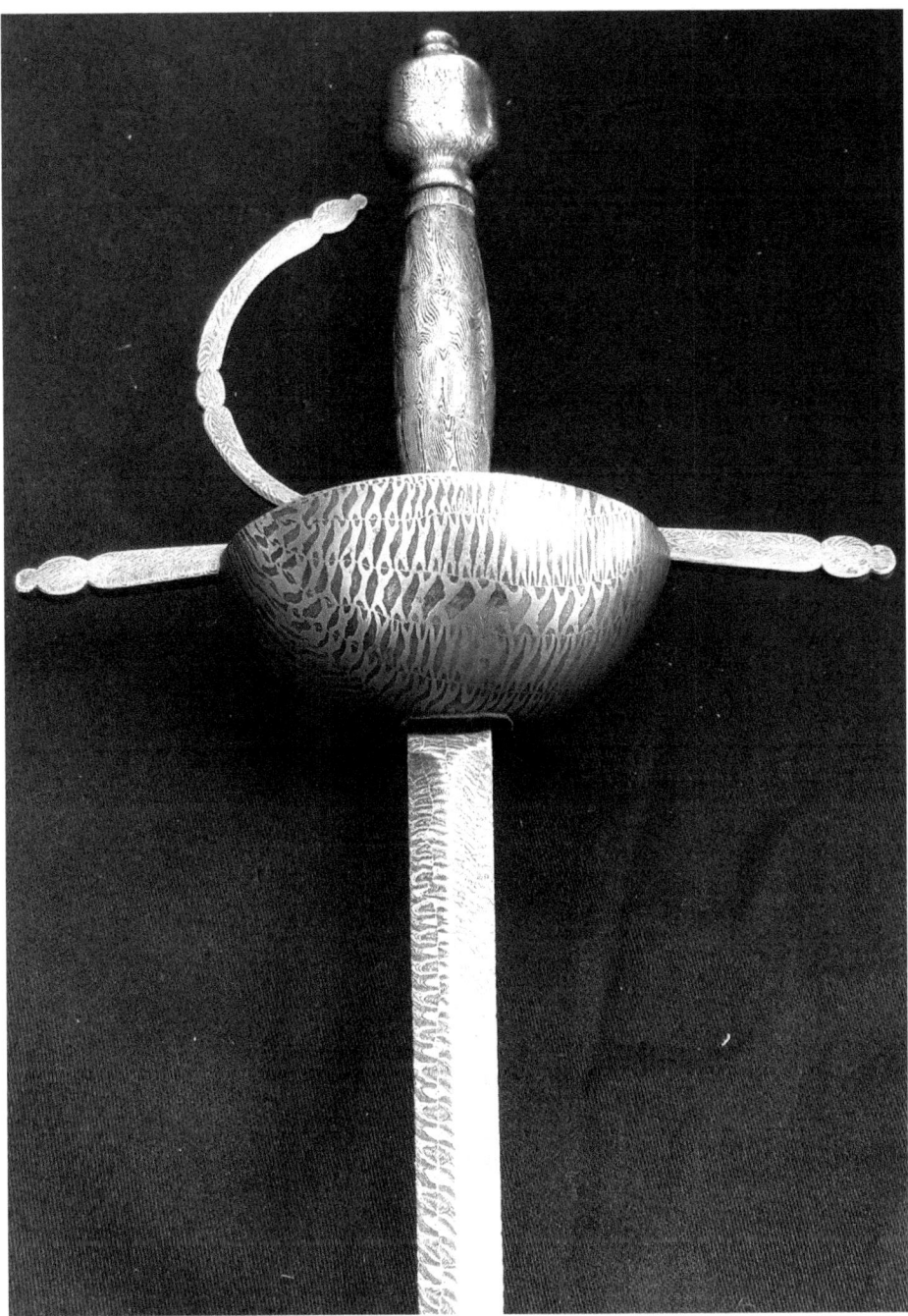

Figure 6.2.2 : Tazza made in Europe in the 17th century. (Total length 112.5 cm). Collection Musée d'Art et d'Industrie, Saint Etienne, Fr.

6.3 The engraved steel imitates and competes with the damasked steel

Chemical engraving, an old technique

Chemical attack makes it possible to produce all kinds of patterns with lower costs. However this engraving is erased with polishing or wear. The process is clearly described by Perret in 1771 [50] *I coated a plate of polished steel with a layer of wax falling from a burning candle and then drew lines on the blade with a steel point, I poured etching on steel, let it bite the exposed part for an hour and then washed the blade. I discovered the secret but it's cheating since the steel is sold as damask steel.*

Mechanical engraving was also practiced to dig grooves by abrasion for receiving the gold inlays, the famous damascening which made Toledo famous. The wootz steel swords often comprise incrustations like that of Figure 6.6.2. This work is delicate because the blade should not be heated.

Gilding

Etching was the preliminary stage of preparing sumptuous decorations of blades of swords around the 18th century. Landrin [52] explains in the manual of the knifemaker how to realize the gold gilt on metals (silver, bronze or iron), that is to realize vermeil. The process consists in applying a mercury and gold alloy which is a very ductile mixture comprising eight parts of mercury for one of gold. *The amalgam of gold is applied and mercury volatilized by exposing the instrument to fire. The gold film remained on metal is then of a brown dark color.* Then follows the receipt of a coating suitable to give the natural color back. Other methods are proposed which consist in soaking the part to be gilded in gold salt solutions, these methods usable only for iron and steels.

Mercury was known to the Chinese since the first millennium BC. Cinnabar, mercury salt was considered a recipe for longevity, if not eternity. The technique of gilding with gold dust and with mercury was used in China in the 4th century BC and propagated to the West in the 3rd and 4th centuries AD. It was known to the Etruscans and Romans (text of Pliny the Elder). The latter technique was abandoned in the 19th century when its toxicity has been proven. From that time dates the case of the gilding of the copper plates of the St. Isaac's Cathedral in Saint Petersbourg where sixty workmen assigned to this work died poisoned by the mercury vapors.

There are two other processes of gilding: gilding by application of gold in the form of very fine gold sheet and gilding by immersion in gold salts with or without electric current. Gilding by electrochemistry is the method which was substituted, it is usable to gild copper, iron and bronze.

At the end of the 18th century, this technique produced magnificent pieces such as the saber of General Kellermann (1735-1820) shown in Figure 6.3.1. At the same time, General Hoche (1768-1797), his junior by 33 years had a damask sword.

6.4 Making gun barrels with damask

The first European rifle barrels were relatively thick for safety and consequently very heavy. It is surprising that the European smiths realize only later that it was possible to make lighter and more reliable firearms by using damasked steel. The example came from the East with the first guns, spoils of war taken to the Turkish army during the Vienna siege in 1683. A few years later during the Egyptian campaign Europeans were faced again with oriental weapons. Therefore, the gun barrels were made with the technique as practiced by blacksmiths to manufacture knives and cutting weapons.

The manufacturing of the damasked artillery experienced considerable development until the end of the 19th century with strong competition between the English, Belgian and French manufacturers. The catalogs proposed hundreds of models. Prizes were awarded to the best realizations, an inscription on the gun says: 2nd prize in 1821 followed by the name of the manufacturer Antoine Merley Chomellon (Figure 6.4.1 A).

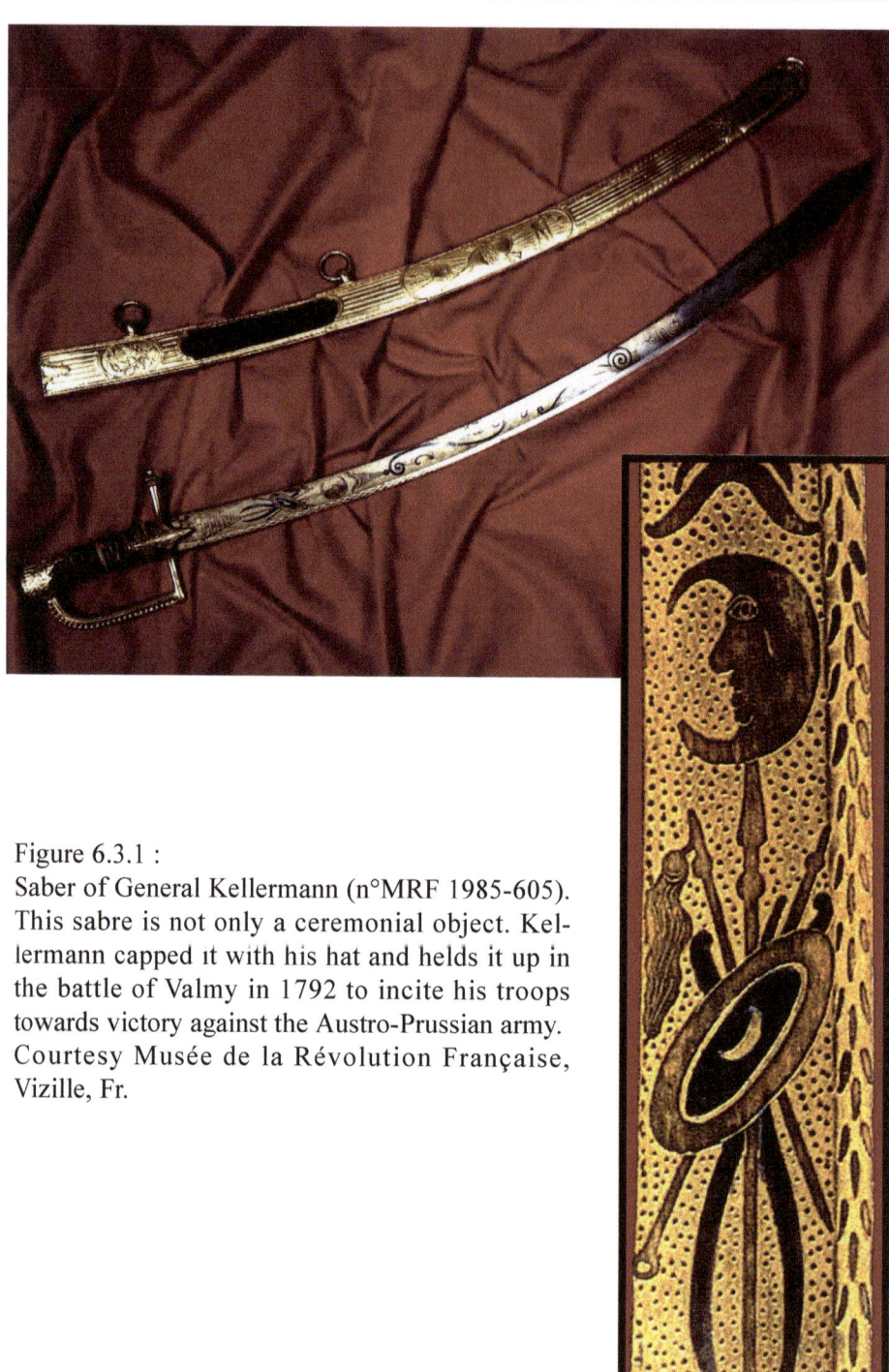

Figure 6.3.1 :
Saber of General Kellermann (n°MRF 1985-605). This sabre is not only a ceremonial object. Kellermann capped it with his hat and helds it up in the battle of Valmy in 1792 to incite his troops towards victory against the Austro-Prussian army. Courtesy Musée de la Révolution Française, Vizille, Fr.

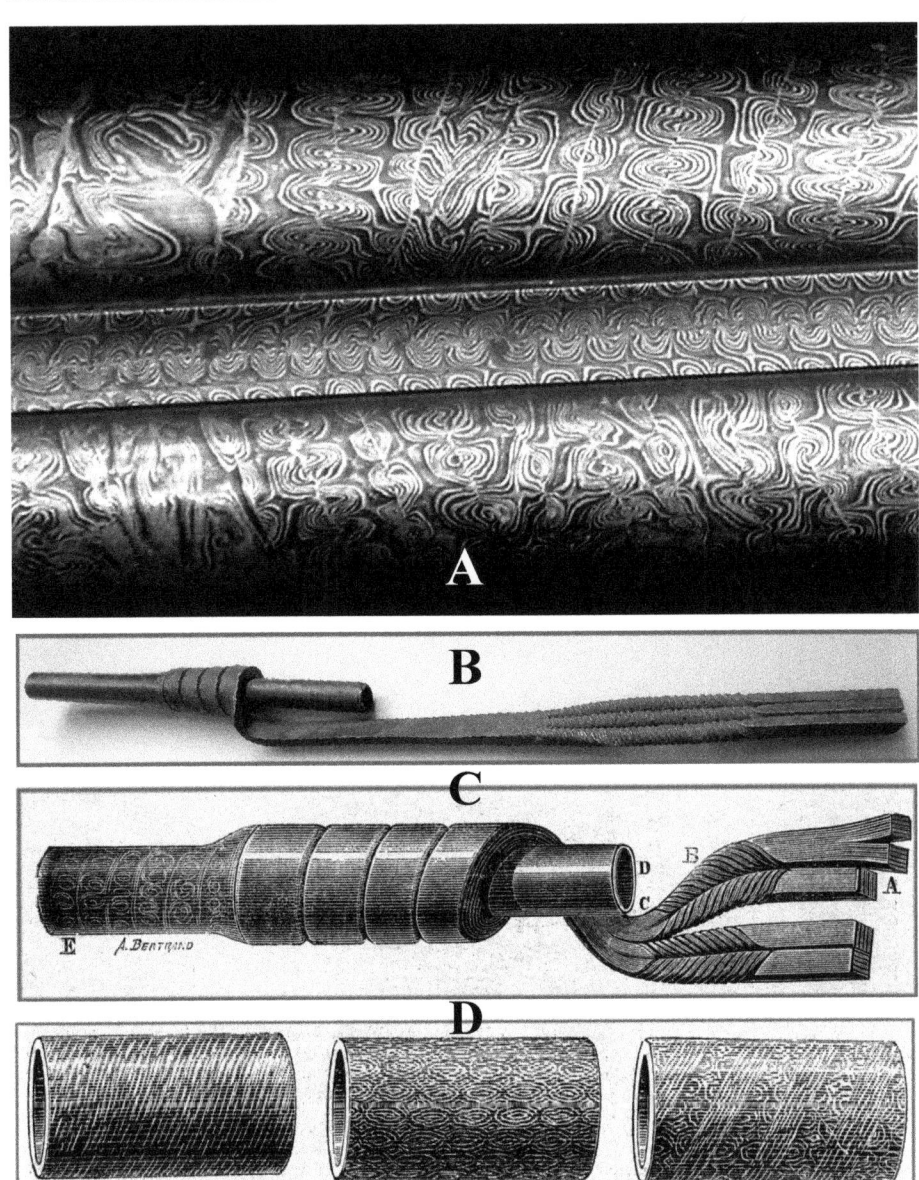

Figure 6.4.1 :
A) Photo of a double barrel shotgun with a beautiful damask pattern.
B) Coils for the manufacture of guns with three rods. C: With four rods
D) Optimizing performance by the pattern of the damask.
A, B) Collection Musée d'Art et d'Industrie de Saint Etienne, Fr.; C, D: Catalogue of the Manufacture Française d'Armes et de Cycles de Saint Etienne

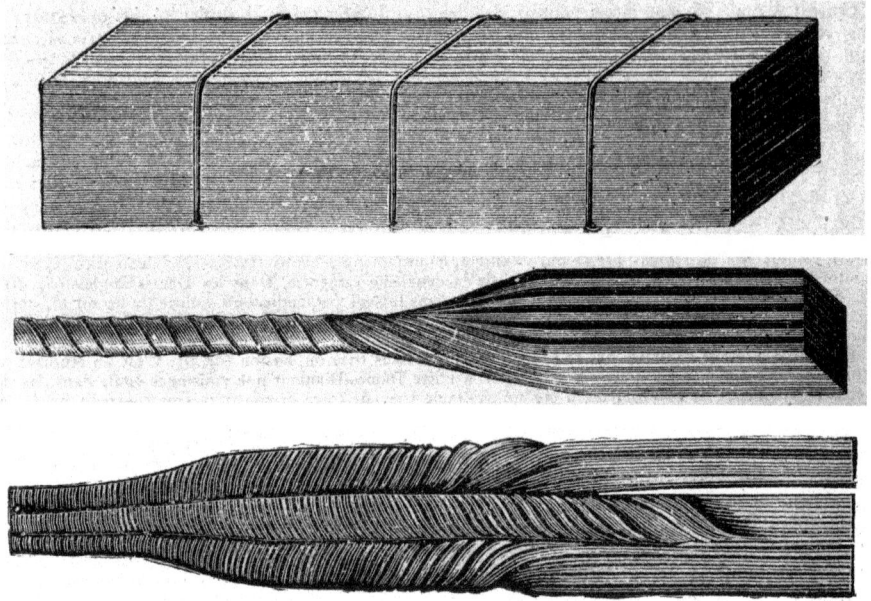

Figure 6.4.2 :
Forming the bundle, twisting the rods, combining the rods.
Catalogue of the Manufacture Française d'Armes et de Cycles de Saint Etienne

The steps in this manufacturing

The following explanations are taken from a catalog of French Manufacture d'Armes Cycles de Saint Etienne in 1898.
- **Preparation of the piece :** gathering the bundles of 50 steel wires and 50 iron wires 1cm square, 50 cm long arranged in a checkerboard pattern (Figure 6.4.2 A).
- **Stretching of the welded bundles** : the pieces are welded into a bar and then rolled and cut into a strip of 12mm in length.
- **Twisting the strips :** white-red (Figure 6.4.1B and C).
- **Assembling the strips :** they are combined by two, three or four to form a steel band.
- **Winding :** the steel band is spiral-wound around the core tube and mandrel.
- **Forge :** the spiral is welded by strong hammering at white-red.
- **Grinding :** the gun is worked outside on grinding stones which turn at the same time as it.

- **Boring**: the gun is finished internally, the core tube disappears.
- **Assembly**: two guns and dividing strips are securely fastened together.
- **Brazing**: the whole is stripped, then topped with copper and wrapped in clay. The whole is heated to white in order to make the copper run into all the parts to be welded.
- **Finish**: polishing inside, planing and outside polishing.

Concern for the mechanical performance

The manufacturing technique by rolling-up has not only an aesthetic objective, but, moreover the production of parts resistant and reliable from a safety point of view. The catalog of the *Manufacture Française d'Armes et de Cycles* presents its new *Damas-Eclair* product in pointing out technical arguments :
- The barrel on the left hand of Figure 6.4.1 D is in *Damas Bernard* or *English*. The threads of iron or steel are transversely arranged in the direction of the gun. This arrangement presents a high transverse resistance and a lesser longitudinal resistance.
- The barrel in the middle is in frizzy Damas or *Boston*. It has, unlike the preceding, a higher resistance longitudinally than transversely.
- The barrel on the right hand is in damask called Damas-Eclair (deposited on November 21st, 1888) which alternates strips of English Damas and strips of Damas Boston. It offers an excellent resistance in both directions.

6.5 The pattern-welded oriental swords

From the first millennium BC the oriental smiths knew how to work smelted iron by forging, currying and welding and, with the western smiths, they improved this technique in the first millennium AD. Examples in Anna Feuerbach's thesis [12] show blades of this last period which are laminated.

In the second millennium AD the currying was widely used to make prestigious cutting weapons and rifle barrels more particularly in the 17th, 18th and 19th centuries.

An example of a cutting weapon is the short sword or the dagger of Figure 6.5.1 with a damasked central part welded on a blade of classic steel. The decoration is said to be onion skin or onion rings. The golden inlays make it a precious object of ornament.

Weapons of the Indonesian peninsula

The Malay Kris is an original, asymmetrical dagger, typical of the Indonesian peninsula, Malaysia, Java and the other areas of the South Seas. According to Indonesian folklore the first ones date to approximately 250 BC. They are sacred weapons, talismans. Moreover the smiths who made them were considered as priests. The representation of friendly divinities (Raksha) decorates the handles, it is made from solid gold or bronzes with precious stones or, more modestly, carved in wood. The blade evokes the mythical snake *Naga*, if the blade is right, it shows the snake at rest and if the blade undulates it shows the crawling snake. The number of undulations has a magic significance. The scalloped blade is made from relatively thick alternate layers of soft iron and meteoritic iron-nickel alloy. The example of Figure 6.5.2 is representative of the Hindu influence during the Majahahit empire from 1378 to 1478 AD, that of Figure 6.5.3 is a Kris of Malaysia of the 18th century.

Use of meteoritic iron and nickel

The steel blade is built by welding layers originally made up of steel and meteoritic iron-nickel. The use of meteoritic iron reinforces the sacred and precious character. The iron alloy and the nickel were replaced by steels of import probably from the 17th century, at the beginning of the German colonization, then, during the twentieth century, by stainless steels, pure nickel and even recycled steels without nickel. The layers seem relatively thicker than those of laminated damask steels and the pattern is quite visible. The use of nickel or nickel-rich steel makes the contrast between the layers quite apparent. On the other hand, as nickel facilitates graphite formation (graphitization), the blades should not be worked for too long because of the risk of losing hardness. The decorations of the blades are very diverse. Sachse mentions a study counting 23 traditional models bearing each one an imaginative name: up-right broom, flower of the nutmeg tree, golden waves, strands of yarn etc. [32].

THE SWORDS OF THE SECOND MILLENNIUM

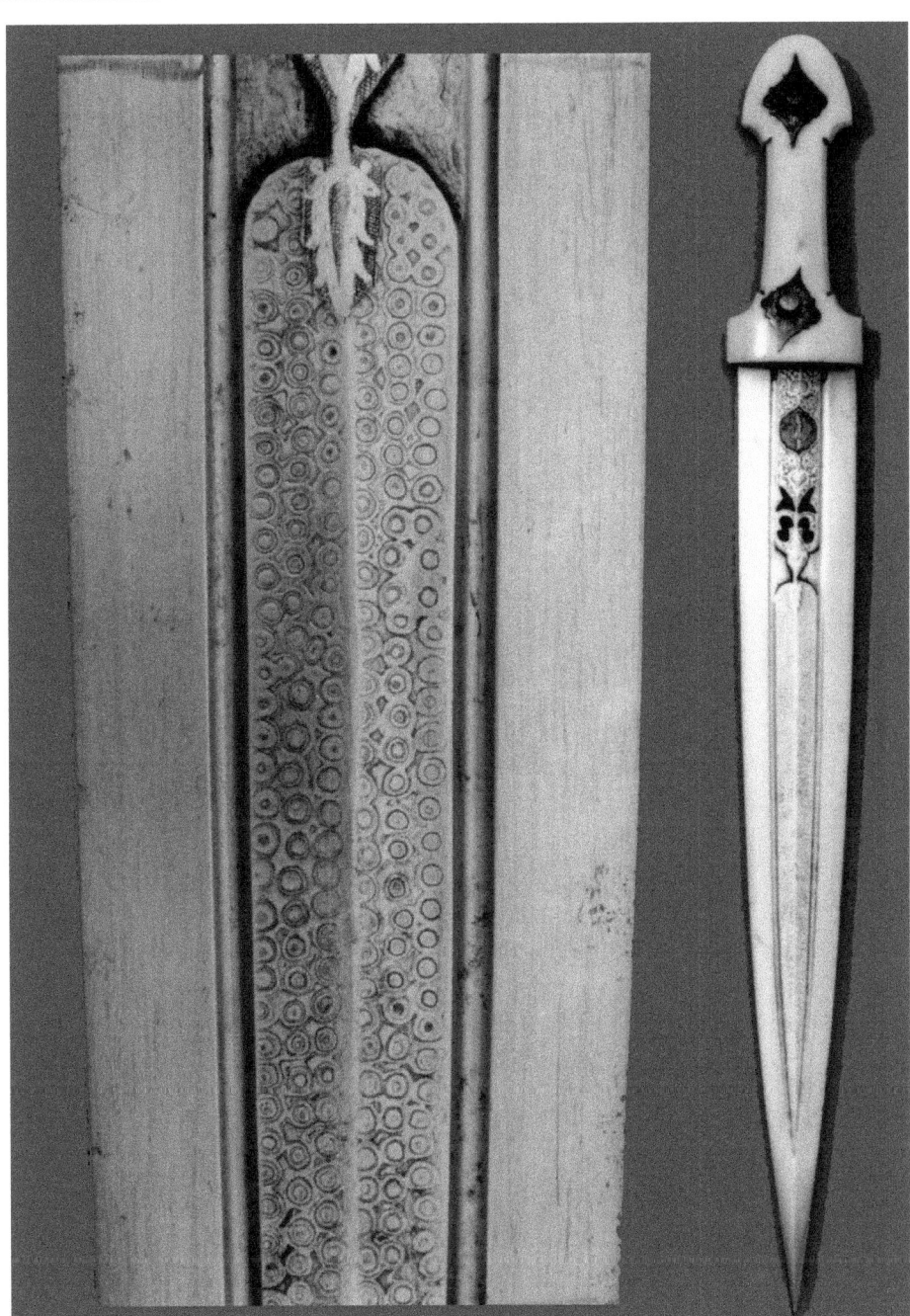

Figure 6.5.1 :
Short pattern-welded sword (51cm) from Iran or Azerbaïjan, 1820-1860, showing an onion ring design. Courtesy Bern Historical Museum, Switzerland.

Figure 6.5.2 : Indonesian Kris from Bali, probably made in the 17th–18th centuries. The handle is massive gold inlaid with gemstones.
Courtesy Bern Historical Museum, Switzerland.

Figure 6.5.3 :
Kris DAPORBENER from Malaysia in the 18th century. Brass handle inlayed with red stones.
Collection Musée d'Art et d'Industrie, Saint Etienne, Fr.

6.6 Swords made with Damascus steel

The forging of crucible steel

In the second millennium, the practice of melting out of a crucible extended beyond Central Asia to the Middle-East, to Ukraine and to the Arabic world, the steel produced being known under various names of *pulad*, *bulat* and *fulah*. The art of forging these steels was held secret, which explains why it propagated very slowly. It took several centuries for it to reach China and Russia in the Middle Ages. It spread south from the Mediterranean to the Arabic world where it was integrated later into the Islamic culture.

Indian steel was used to forge blades of swords considered for their exceptional quality, being lightweight, highly resistant, and with the famous sharp edge *that can halve a silk handkerchief on the fly*. Landrin [52] describes them *elastic, sound, with an excellent sharp edge and with a remarkable delicacy of material. They are characterized by moire designs, a sort of sandblasted black and white, crystallized, which suggests that indeed these blades are made by the meeting of various fabrics of steel, to various degrees of steeling*. The famous decoration is visible on the blades of Figure 6.6.2.

Splendid parts were forged starting from the 13th century, all over the Islamic world including Turkey, Central Asia and Islamic India. These swords with the precious ornaments have created the legend of the Damascus swords.

The European blacksmiths spent several centuries before being able to understand and control the working of Indian steel, so different from their usual techniques: *The knifemaker and furbisher said they would have found it detestable (the wootz) if the celebrity of this steel, acquired by some centuries of an indisputable experience, had not held them in respect*. This steel crumbles when it is forged too hot and becomes brittle when it is forged too cold. Finally it loses its properties when it is forged for a too long time because it graphitizes.

Zschokke explained it in 1924

Zschokke, by microscopic observation, showed that the structure of wootz comes from the alignment of carbides in periodic rows [56], (Fig. 6.6.2). A few decades later, several researchers studied the metallurgical structure of the various types of forged damask blades [26-57]. And recently the metallurgists clarified the genesis of the microstructure of wootz [58-59], (§ 10.2). The analyses confirmed that it was indeed a carbon-rich steel (Table 6.6.1). I

Table 6.6.1 : Range of composition determined on various samples of wootz steel [60]

	C	Mn	Si	S	P	Cu	Cr	Ni
min	1.34	0.005	0.005	0.007	0.05	0.04	trace	0.008
max	1.87	0.14	0.11	0.038	0.206	0.06		0.016

There is a 70 year gap between the modern findings and those published by Zschokke, [56] : *above all, it is necessary to specify that the old genuine damasked steel of India named pulat is not at all some welded steel... We rather have to deal, on the contrary, with steel melted in a crucible, regular and the particular structure of which results from phenomena of crystallization and from segregation.*
Besides of the pattern of damasked steel, by the repeated forging, the ancient Armorers of India and Persia had for main purpose, much less the production of a damascene and the resulting decorative effect than obtaining a metal with a good toughness. This last hypothesis is also consistent with an observation of Belayev, whereby the astonishing beauty of the Indian steel was only a secondary objective and accidental result.

The fashion of the damask structures culminated in the 18th century, at the time where Europeans knew how to forge it correctly. Latest damask swords were crafted in the early 19th century, at the same time as techniques were developed for obtaining high-performance steels. Moreover, at the end of the 19th century, the interest ceased for the swords as weapons. The most beautiful specimens became collectables. It is necessary to pay tribute to several collectors, in particular Moser who gathered a magnificent, still grouped, collection to which belong the specimens of Figure 6.6.2. The Moser collection contains precious cutting weapons and remarkable examples of armors [47]. Figiel's book also presents numerous photos of parts scattered in private homes or in museums [16].

Figure 6.6.2 :
Left hand) Iranien *shamshir* saber 1820-1860, 98 cm long, with a walrus ivory handle. - Right hand) Indo-Persan saber *shamshir*, 93 cm long, bearing the inscription "By order of King Naser, 1165" (Arabian calendar *i.e.* 1738), handle decorated with gold and enamel [47]. Document from the Henri Moser Charlottenfels collection. Courtesy Bern Historical Museum, Switzerland.

The so-called lost expertise

The legend of the lost know-how is ancient. Already Landrin wrote in 1835 [52]: *the Persians made for a long time a mystery of the manufacturing of these blades, of which the most estimated came from the city of Damascus. It even seems that the secrecy of old manufactures which provided all East was lost in the country, since today the Persian blades are of an extreme brittleness and that they were, on the contrary, formerly famous for their elasticity.* Certain authors attribute this change to the exhaustion of the ore containing vanadium, depriving the steel of its magic additive (§10.2). It also seems possible that the quality of these blades fell owing to the fact that they were produced in a more massive, cheaper and faster way to answer the high demand.

The typical patterns

M. Khorasani reports that good quality Persian shamshir blades were made of crucible or wootz steel. Since the 11th century AD, various patterns of steel were produced and distinguished by evocative names: *payhaye murche* (the pattern looks like blazing ants' feet), *moje darya* (sea waves), and *pare magas* (fly's wings), [48-49].

Several authors, among whom Ernst Kläy [47], Manfred Sachse [32] and Anna Feuerbach [12], relate the distinction of five typical patterns on forged blades that are known as *stripey, water, wavy, mottled and mottled woodgrain*. This classification which is also purely decorative seems to have been originally published in a catalog of the Moser's collection in 1955, [61].

All the patterns, except the one named stripey are obtained with high carbon steels. They are formed by alignments of carbides of some micrometers in size, skillfully highlighted in a matrix with a very fine structure (Figure 6.6.2). The periodicity of the carbide alignments produces a more or less disturbed undulating design. The pattern may be deliberately punctuated by expert trips and tricks during hot hammering. These are concentric circles called roses and bands in which the spacing of alignments is tightened to form the pattern variously known as *Kirk Narduban, Mahomet's ladder*, the *Ladder of the Prophet, Jacob's ladder* or the *Forty Steps*).

The designs called *stripey* pattern and sometimes also the one called *water* are observed in the case of low carbon steels and are defined as soft damask. They are found especially in some swords from Syria. The design is formed by alternation of ferrite and pearlite bands. The color of the pattern arises softer, less contrasted than that obtained starting from steels with high carbon content. From the metallurgical point of view the involved mechanisms are very different and will be analyzed in detail in chapters 13 and 11.

6.7 Japanese swords

Swords in the history of Japan

There is an abundance of literature published on the Internet concerning Japanese swords which can be found in many well documented sites: Wikipedia, Museum sites, commercial sites and also personal sites of enthusiastic amateurs. The subjects are mainly classification according to their shapes, lenghts, places of origin, school of sword-making or their history. Consistent with these information sources, a short summary of the history of antique swords is proposed to illustrate the place of sword-making in the Japanese culture.

The oldest swords discovered in Japan go back to the 3rd century of our era, they are of Japanese origin. The techniques were then imported from China via Korea. By the 5th century, straight single-edged swords, called *chokotu* were manufactured in Japan. It was during the Heian period (794–1184), when the capital was established in Kyoto, that the specifically Japanese technique developed. The shape of the sword called *tachi* became bent, doubtless to adapt itself to new techniques of fight and also thanks to the technical advances of smithing. The structure of the sword was already made of a ductile core and an outer shell at the surface that is richer in carbon, and as a consequence harder, at the cutting edge of the blade. Afterwards, every period produced very specific swords, defined by their length, width and curvature.

The next period dated from 1184 to 1333 and is known as the Kamakura period, because, although the emperor remained in Kyoto, the government was run from Kamakura, which became the cultural capital. At the end of this period, two Mongolian invasions were pushed back in 1274 and 1281. Fights always reveal weaknesses of weapons on each side. Also, thereafter *tachi* swords were strengthened.

During the Nambokucho period from 1334 to 1393, the heavy, long, two handed swords or *odachi* appeared. They were accompanied by one second sword the *katana*. During the Muromachi period which extends from 1394 till 1595, the heavy swords disappeared and were replaced by the *katana*. Very nice blades were forged early in this period, it is the great time of the katana which was very appreciated in Japan but also in China where it was exported. Then the internal conflicts and rebellions increased the demand with the consequence of a production at lower quality.

The Edo period from 1596 to 1867 was a long peaceful period in which a new organization of Japanese society settled. Edo was the political capital and Osaka an important commercial place.

In Edo (Tokyo nowadays) samurai exhibited their power at the court of the governor with luxurious swords. But also in Osaka, the merchants, who were authorized to wear short weapons, paraded with expensive swords. Even the actors wore weapons, the two-edged sword being the attribute of the bandit. Figure 6.7.1 is a sketch according to a print dated in the region of 1782 inspired by the Kabuki theater. The frozen attitude, the costumes of the characters, and the conventional attributes are characteristic of this period.

The art of making swords or daggers reached a very high degree of sophistication. The blades of the swords were adapted to a practice of advanced fencing with a stabbing weapon.

In Japan the warriors were called samourai. It is at the beginning of the 11th century that they began to constitute a power, a dominant caste. They kept this power until 1868.

The samurai were valiant, brave men who had a very strict honor code. They were cultured, art lovers in all its forms. Their culture was in accordance with the religious thought of their period, in particular in the 13th century when a form of Buddhism appeared, the Zen Buddhism, which required a severe spiritual and body discipline. The influence of the samurai in Japanese culture lasted until 1867, the year of the abolition of the samurai class.

Figure 6.7.1 : Sketch from a print MM.

The revival of an art

As in many countries, in the 19th century the sword lost its utility as a weapon. Already the disappearance of the samurai and the prohibition of the bearing of sword in public had made it lose its representative role. Nevertheless, the art to make swords survived, as it survived numerous difficult periods in Japanese history. The latest was from 1945 to 1953, after the Second World War, during which the manufacturing and the possession of weapons were prohibited to the Japanese. Most of the craftsmen converted to other activities. However, at the conclusion of this period there was a real revival of this art. Some blacksmiths revived the traditional art, such as in Kyoto Yoshindo Yoshihara, a worthy descendant of ten generations of swordsmiths [62-63]. The enthusiasm was such that the government has feared the appearition of a mass production without real quality and respect of the tradition.

A company founded in 1960, the NBTHK (in English Society for the Preservation of Japanese Art Swords), created multiple activities to control the training of the apprentices, to link the smiths, to organize competitions, to deliver certification, to establish a classification etc. A blacksmith arrived in the forefront of the rankings for several years and is recognized by the Ministry of Culture as a *Living National Treasury*.

The bibliography on the ancient and modern Japanese swords is rich in books and also more recently in numerous web sites of which that of the Kyoto National Museum particularly provides images of ancient swords. Fascination for Japanese swords is always vivid if one considers the great number of personal sites very well documented.

The structure of Japanese swords

The main features are illustrated by two modern Japanese swords (Figure 6.7.2), *katana*, made according to a strictly codified traditional procedure. The large edge of the blade in clear contrast is the hamon, the harder part of the cutting edge, which after polishing becomes brighter.

The *hira-tsukuri* style blade shows a hamon which appears light, with a wavy and loop pattern called *choji midare*. The blade is also decorated by traditional *horimono* engravings. The *shinogi-tsukuri* style blade is

Figure 6.7.2 : Two modern katana (75 cm length) made by Yoshindo Yoshihara in Kyoto.

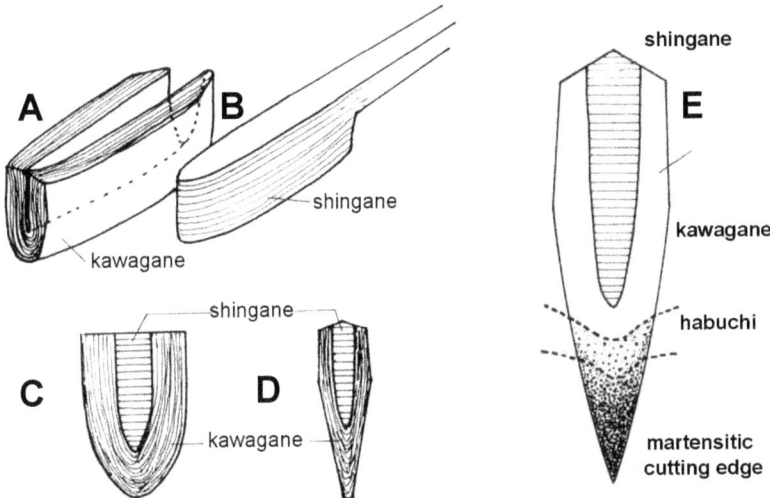

Figure 6.7.3 :
Fabrication process of a Japanese sword, according to [62]:
A) Preparation of the outer envelope, made from *kawagane* steel (medium steel);
B) Preparation of the core, made from *shingane* steel (soft steel);
C) Forge welding of the two parts;
D) Forging of the blade;
E) Details of the cutting edge or *hamon*. Schematic transverse section, showing the region of martensitic structure at the cutting edge. The intermediate region, or habuchi, shows patterns such as *nie* or *nioi*.

so named because of the ridge along the blade on both sides. The *hamon* is again a *choji* type pattern with characteristic loops, wider at the top than at the base. The two blades were made by Yoshindo Yoshihara, using the traditional Japanese process.

Blade construction

The blades are usually made from two steels, one moderately carburized, the *kawagane* and the other one weakly carburized, the *shingane*. The traditional starting material was what known as *tamahagane*, a strongly hypereutectoid steel (1.2-1.7% C) obtained from soft iron by carburizing with charcoal, sometimes replaced by imported Indian wootz steel.

The core structure is made out of ductile steel *shingane* which gives a great flexibility to the blade. This steel (Fig. 6.7.3), is obtained by a large number of repeated forging and folding cycles. During these hot working operations inclusions are expelled and carbon gradually removed by oxidation at the surface leading to a practically uniform carbon content of about 0.2 %, which ensures the required ductility.

For the outer envelope, the starting metal is selected from among the high carbon steels, possibly Indian steel. *Kawagane* steel is prepared in the same way as the *shingane* steel by a succession of forgings and foldings which are less numerous in this case and the loss of carbon is limited, with a final average level of about 0.5 %. The diagram in Figure 6.7.3 according to [62] presents the assembly which is designed to combine two qualities of steels, a harder outer jacket of steel wrapped around a softer inner core of steel. Other types of assembly are performed, some are similar to those of the steel *sandwich* cited § 9.1.

The important point is that the weld zone does not coincide with the limit of the hardened area. Indeed, when the blade edge is partially hardened, a huge internal stress is exerted due to the formation of martensite. Induced expansion gives the final curve to the blade during quenching. If the area adjacent to the martensite is not ductile, for example with the presence of a weld, there is shear failure.

The hamon

The most spectacular feature of Japanese swords is the martensitic structure of the cutting edge, known as the *hamon,* shown on Figures 6.7.2 and 6.7.4, which was obtained by specific treatment. Various clays in different thicknesses, are coated on the blade. After austenitizing, the coated blade is quenched. Only the edge being left bare is exposed directly to the water, and thus the cooling rate in this region is sufficient to locally induce martensite transformation. The clay coatings were applied in such a way that the transitions in microstructure either occurred along straight lines, or in the form of patterns or silhouettes (Figure 6.7.4). The design is emphasized by a careful polishing and etching. However, in this case, it is relatively easy to reveal because the martensite is very hard, much harder than the rest of the blade.

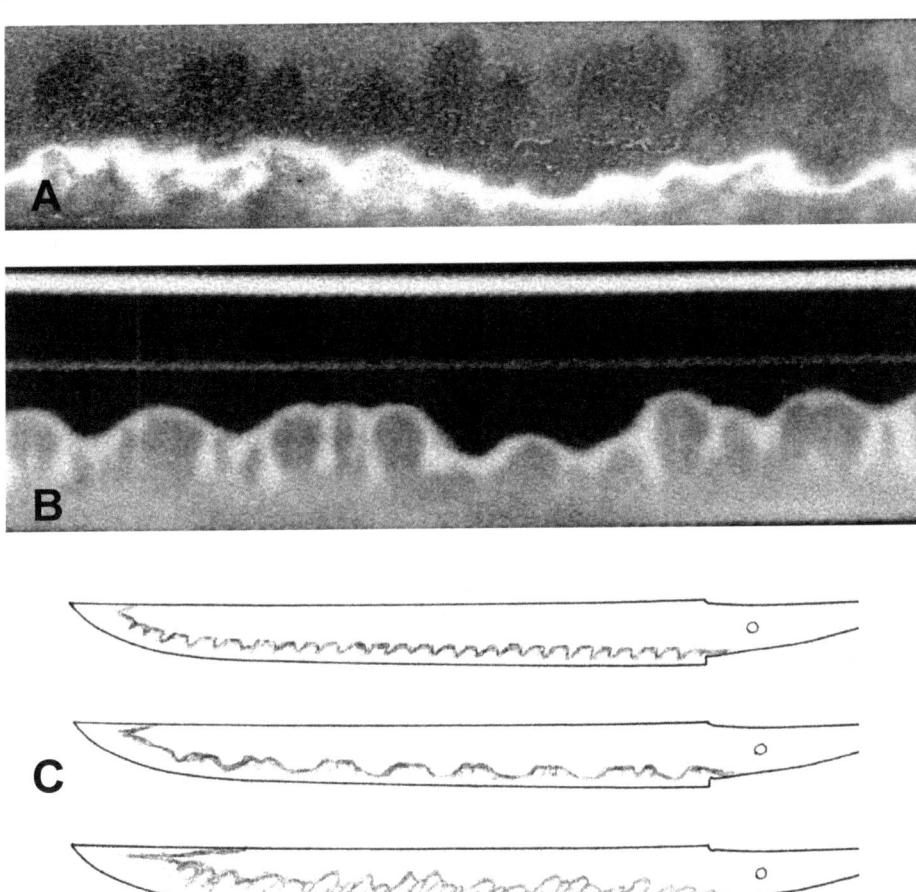

Figure 6.7.4 : The hamon
A) Longitudinal close-up of a 14th century sword from Bizen (Okayama). The *hamon* design is called *gunome midare*. The clear zone is martensite, the two dark zones, *i.e.* the main blade surface, are a mixture of ferrite and pearlite or *nioi*, in between the dark and the clear zones it is composed of fine ferrite and bainite. The granite-like appearance of this part of the blade is due to the variations in carbon content in the different layers produced by the folding and hammering process (see §11.2). The grey zone opposite to hamon, is called *utsuri*, meaning *mirror image of the hamon*, and is characteristic of swords of this period, probably being caused by non-uniform heating of the blade.
B) *Choji-midare* hamon pattern on a modern sword. The white line corresponds to a change in slope on the blade surface.
All photographs by courtesy of Yoshindo Yoshihara, Kyoto
C) Schematic hamons after [62].

The ji-hada

Hada means the pattern on the surface of the blade, it is revealed by slight etching. This pattern depends upon the way the steel was forged, and the forgewelded layers were folded. The design shown in Figure 6.7.4 A is of *itame-hada* type, a name which evokes the look of wood grain.

In fact, different repartition of ferrite grains and pearlite cells are thus observed (see §14.2). They are conventional microstructures of carbon steels, but a poetic vocabulary was coined to describe this appearance, with expressions such as *nie*, a fine dispersion of silvery sand, and *nioi*, a multitude of cherry blossoms palely lit by the rising sun.

The design was not apparent in the first swords. The development of the decoration was done later by a sophisticated polish. Many different codified procedures were developed to obtain various effects defined in great detail by the choice of various etchants, of abrasives and alternating multiple sequences etching and polishing. The difficulty of polishing *hada* results from the fact that ferrite and pearlite are relatively ductile. The apprentice spends one year just to learn polishing of the blades. The result is the exceptional aspect of the surface.

Numerous metallurgists have admired the immense skill involved, pointing out the almost ritual respect of tradition, namely Bain [64], Smith [26], Tanimura [65], Inoue [66-67].

7 Art and Technology in the third millennium

7.1 The Renaissance of an art and craft

After World War II, stainless steel replaced silver plated copper-nickel-zinc alloy for the applications of usual cutlery. Knives, spoons and forks were industrially manufactured. Concurrently to this mass production, the craft manufacture of personal knives was developed. It was the knife of the hunter, of the fisherman and it became a collector's knife. The blacksmiths realised true artistic works. Among these creative blacksmiths, I will particularly mention Manfred Sachse who played a major role during the second half of the 20th century in rehabilitating damask steel (Figure 7.1.1). Beside his own esteemed artistic

Figure 7.1.1 : Manfred Sachse in his workshop

production, he traveled all over the world to discover the old methods of forging, to reconstitute the manufacturing of many old parts. His findings are presented in his book *Damascus steel* [32].

Within the framework of this book, only some achievements are presented to illustrate the various types of manufacturing. The reader will be undoubtedly surprised to recognize all the old processes used by the blacksmiths to manufacture the blades of swords. The implementation is however facilitated by the use of machines or modern devices such as electric furnaces, forging presses, rolling mills, cutting machines by electron discharge machining etc. Two processes are the result of modern high technology which use powder metallurgy and laser remelting. However, a few blacksmiths still keep a traditional approach with charcoal furnaces and a mass to hammer.

7.2 The search for the best cutting edge

The buyer of knife demands an excellent cutting edge for his blade and this criterion is decisive for the choice of steel grade. Various options are available with advantages and disadvantages. Apart from the properties of the steel, the cutting quality depends greatly on the shape given to the blade edge while sharpening.

The knife which rusts has the reputation to cut well!

An ultra low carbon steel or pure iron can be very easily sharpened and a good sharp edge acquited. Regrettably, they are ductile and quickly lose their edge by wear resulting from use. A medium carbon steel become very hard after appropriate heat treatments which induce its transformation into martensite and it is more difficult to sharpen but highly resistant to wear. Its major drawback is the low corrosion resistance. A good compromise is the choice of stainless steel grades known as martensitic (see § 8.7) which may also acquire a good cutting edge.

The austenitic stainless steels are too ductile to maintain a sharp edge in use. On the other hand it is possible to give them a shiny surface polish, a high quality mirror finish. This quality makes them useful for food tanks and containers because of the facility of cleaning.

Figure 7.2.1 :
Sharpening. Image from the catalog of the "Manufacture Française d'Armes et Cycles" published in 1928.

The micro cutting edge of very high-carbon steels

The excess carbon present in steel is transformed into very hard carbides. When these carbides are very small and regularly distributed in the matrix, they work as micrometer size teeth of a saw. They are very resistant to wear insofar as the matrix is able to encase them firmly. To maximize the proportion of carbides the tendency is to select high carbon steels. Such steels require a specific preparation starting from metallic powders to obtain a fine distribution of carbides (§ 8.10) and appropriate heat treatments for a complete transformation of the martensite matrix.

Quenching to selectively harden the sharp edge

Ancient steels used for the cutlery industry are low alloy steels and thus very sensitive to cooling rate. Blacksmiths took advantage of this sensitivity to selectively quench a zone at the level of the sharp edge. For this purpose a coating deposited on a specific area acts as a thermal barrier slowing down the cooling rate of parts of the blade during quenching (Figure 7.2.2).

The sharp edge of the blade remaining naked is quickly cooled and is transformed into martensite. This zone can be delimited so as to create

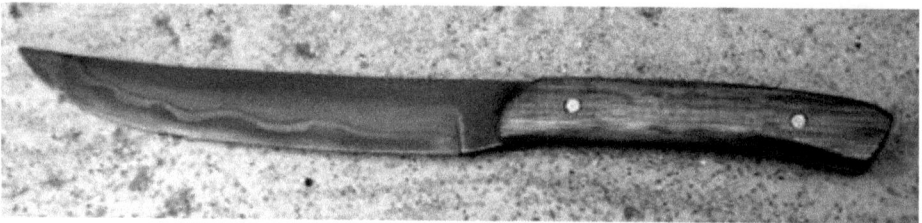

Figure 7.2.2 :
Blade coating before quenching and knife with genuine hamon.
Blade steel is 100Cr6. The coating is a mixture of sand, clay and charcoal in three equal parts. The mixture is reduced into dry powder, then water is added to form a plaster to be spread over the blade. The coated blade was heated and then dipped in warm salt water.
Courtesy Sébastien Masson, Forges de Grésigne, Fr.

a decoration as in the case of the *hamon* of the Japanese swords. Selective quenching is performed successfully with modern steels such as the steel 100Cr6 example. This steel used to manufacture bearings develops excellent properties of hardness and wear resistance after suitable heat treatments. However, the formation of the *hamon* is possible only for some steels.

Martensitic areas are very hard while the pearlitic areas are soft, they show different gloss after polishing. Besides, they react differently to the chemical etchants, which makes it possible to highlight the transition by light chemical etching. Thus, the hamon constitutes an element of the decoration of the blade.

The martenitic transformation corresponds globally to an expansion which is more significant for high carbon content (only carbon dissolved in austenite solid solution should be taken into account *i.e.* excluding carbon of carbides). Strong internal stresses are generated which may result in a curvature of the blade. Though many smiths have been disappointed to obtain a reverse curvature to what should be expected after quenching (see discussion § 8.2).

Purely decorative fake hamon can be simulated on any steel by a selective etching or by grinding the zone of the cutting edge. The aesthetic result is perfect in the case of etching, however, polishing or simply wear make superficial decoration disappear.

Sandwich steels with a strengthened cutting edge

Creating a composite material is often a solution to obtaining a material with enhanced properties, better adapted than any of its component materials. This is the case for blades with three layers of steel. The combinations of grades are numerous but they always associate a hard grade with a core which appears on the sharp edge and a more ductile one which forms the surface of the blade on both sides.

The first example is a hand forged blade (Figure 7.2.3). Its conception answers a threefold objective: a highly efficient cutting edge, a decorative surface appearance and excellent mechanical strength and good toughness especially for the bending test. The sharp edge is constituted by 1.1-1.4 %C steel to which the specific heat treatments confer an exceptional hardness while preserving a certain ductility (see § 8.4). The second grade is an ancient soft iron, that is to say puddled iron without carbon (such as the one used for the Eiffel Tower). A weak hardening is brought about by the presence of a low phosphorus content. The zone of the cutting edge is transformed into martensite during the water quench whereas the coating of the mild steel is not transformed. The result is somewhat similar to the hamon but it is not a hamon as the local hardening is due to the difference in composition. For aesthetics, the part near the back of the blade is finished with only a slight rough-hewing.

The second example is a three layer bar of industrial production intended for the manufacture of blades. The two steel grades are chosen in the range of stainless steel: a martensitic steel with 0.68 %C and 14 % Cr

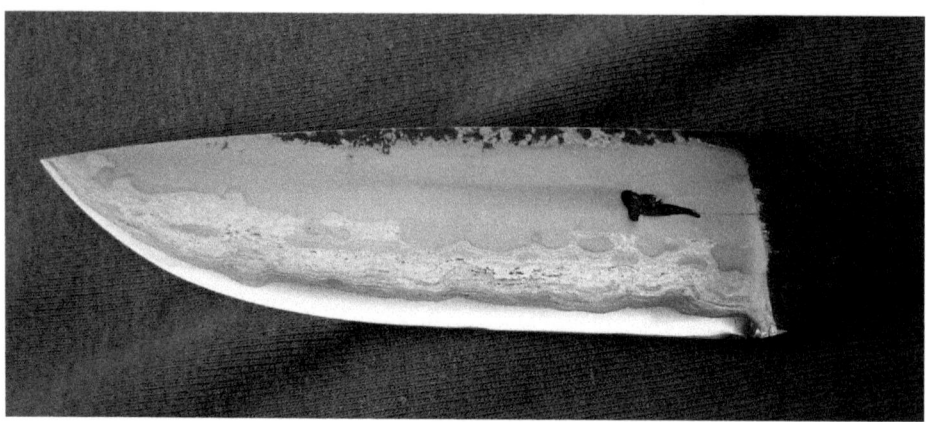

Figure 7.2.3 :
Knife blade with three layers, a core layer made of high carbon steel, wrapped between two layers of soft iron; the transition zone exhibits a decoration.
Knife forged by Eric Plazen, Montjoie, Fr.

for the thinned zone on the cutting edge which is wrapped between two layers of 18/8 type austenitic steel.

The micrographs in Figure 7.2.4 show the junction zone between the two steels marked out by a string of very small globular precipitates of oxides indicated by the arrows. A strip of about fifty micrometers between the weld and the high-carbon steel appears deprived of carbide. It is the transition zone in which carbon has diffused from the high carbon steel to the low carbon steel. Taking into account the diffusion rate of carbon, this zone may be considered as large, thus revealing a high temperature in the welding process. From a mechanical point of view, the progressive transition between grades accomodates the poor compatibility between the two steels, thus enabling a good resistance to rupture. In this case the transition between the two zones is not intended to be a decoration on the blade, it is hardly detected by a slight difference of brightness between the two steel grades after polishing.

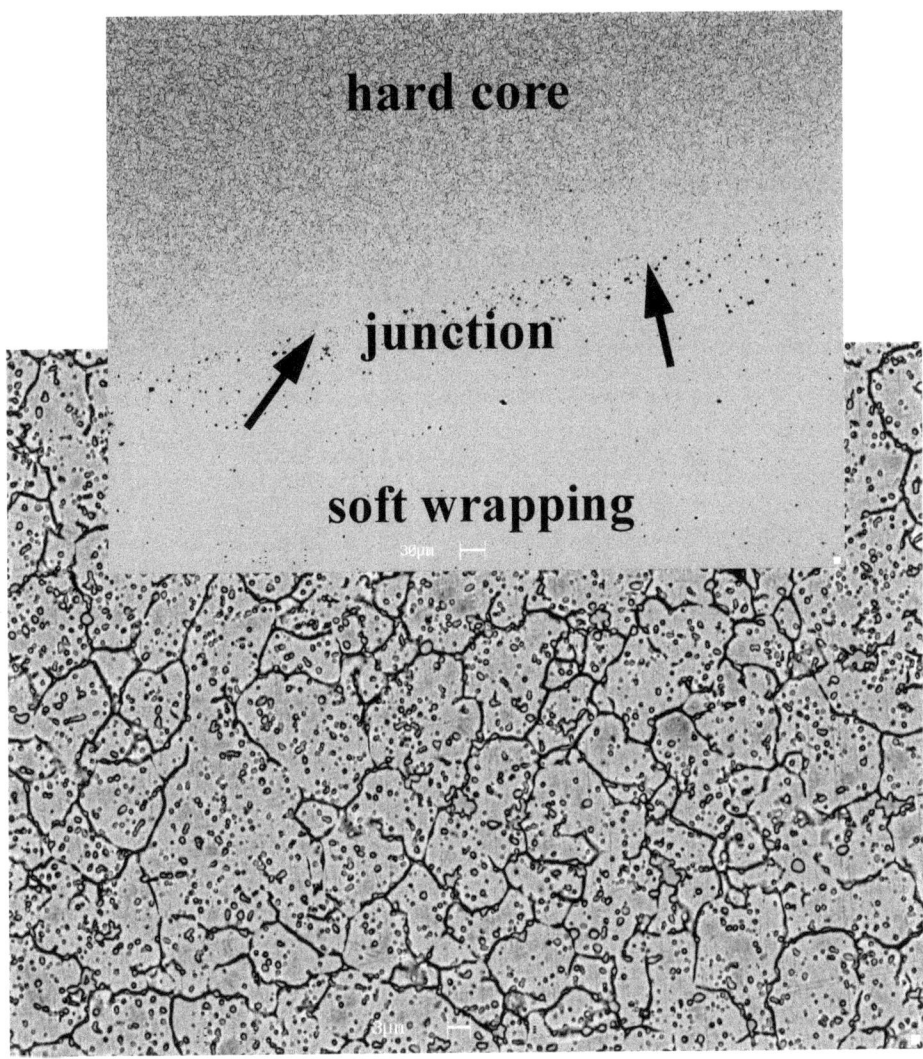

Figure 7.2.4 :
Electron micrographs of a sample air quenched from 1060 °C then held at 200 °C for an hour (etchant: 15 ml HF, 4 ml HNO_3, 80 ml ethanol).
Top : hard core in T7MO steel,
0.68C-0.64Si-0.001S-0.016P-0.43Mn-0.15Ni-14.24Cr 0.55Mo
and soft wrapping in Dauphinox A3
0.02C-0.39Si-0.022S-0.026P-1.05Mn-8.05Ni-18.21Cr
Bottom : magnification of the hard core
Courtesy Aciéries de Bonpertuis, Fr. ; Document Grenoble INP, Fr.

7.3 Multilayer steels

Pattern welding

Most of the pattern-welded blades are made with two grades of steel selected to show the best contrast and the best glitter after etching. In Table 7.3.1 a few grades are classified in two series: one of shiny steels and one of darker steels. Other properties must be matched to choose a couple of grades, - melting points, since welding is difficult when they are too different and, - thermal expansion which may break the weld between the layers when the effects are too different.

The layers are folded up and forged several times [68]. A good contrast is achieved with low carbon steels containing manganese, associated with medium carbon steels typically used in conventional cutlery. A small addition of manganese promotes the formation of carbides and prevents the formation of graphite when forging lasts too long. An addition of chromium inhibits the diffusion of carbon between the layers and prevents the homogenization between them [69].

A decoration can be created by incorporation of pure metals. Nickel and pure iron are often chosen because of their bright silvery sheen, but they are ductile and do not perform as well as steel for cutlery applications. Cobalt is forsaken in spite of its beautiful bronzed color. Platinum was used in the 19th century but it is obviously too expensive for common use. Copper and gold are problematic because the melting points are far from that of iron.

Nickel does not form carbides and inversely promotes the transformation of the carbon of the steel into graphite. A smith had noticed that the co-forging of steel and meteoritic iron which contains nickel gave an increasingly soft product; in fact steel was transformed into graphite and ductile ferrite. It is said that nickel is graphitizing whereas manganese is antigraphitizing. Other effects of alloying elements are to be considered such as their ability to facilitate or inhibit the martensitic transformation.

Stainless steels, although considered by the blacksmiths more difficult to forge weld, are nevertheless associated for the manufacture of damasked steel. This is the case for the paper-knives presented in Figure 7.3.2. made out three grades. The core is very hard, made from

Table 7.3.1 : Coupling of steels

	Steel grade	Composition (weight%)
Dark	1084	Fe-0.80C - 0.60Mn - 0.040 P - 0.050 S
	1095	Fe-(0.9-1.03)C-(0.3-0.5Mn)-0.040P-0.050S
	5160	Fe-(0.560-0.640)C-(0.75-1)Mn-(0.7-0.9)Cr-(0.150-0.3)Si-0.035 P- 0.04S
	52100	Fe-(0.98-1.10)C-(1.30-1.60)Cr-(0.25-0.45)Mn-(0.15-0.30)Si-0.025P-0.04S
	20C	Fe-1.00C-0.30Si-0.45Mn-0.15Cr-0.015P-0.0005S
	90MCV8	Fe-0.8C-2Mn-0.4Cr-0.10V
	W-2	Fe-(0.6-1.40)C-0.25V
Bright	L6	Fe-0.7C-0.5Si-1.75Ni-0.5Mo-0.25V-0.25Cu
	15N20	Fe-0.75C-0.30Si-0.40Mn-0.11Cr-2Ni-0.018P-0.005S
	45NCD16	Fe-0.4C-0.5Si-(0.2-0.5)Mn-1.2Cr-(0.15-0.35)Mo-(3.8-4.3)Ni-0.03S-0.03P
	ASTM 203 E	Fe-0.23C-0.8Mn-(3.25-3.75)Ni-(0.15-0.30)Si-0.035P-0.04S
	Nickel	

steel prepared by powder metallurgy. It results in a decorative effect of relief on account of the different hardnesses of the layers, thus even wear preserves and enhances this effect. The folding knife in Figure 7.3.3 was made by machining pre-damasked stainless steels.

Figure 7.3.4 shows another example of a blade with a hard core wrapped in pattern welded steel (sandwich steel). Hard steels liable to develop an excellent cutting edge are usually chosen for the cores.

Figures 7.3.5, 7.3.6 and 7.3.7 show knives with different types of bolsters. The bolsters are designed to strengthen the blade/handle junction and the butt of the handle. It is also a decoration, having an embellishment with engraving, inlay of materials and sometimes gemstones. The steel grades may be chosen from a large selection since this part does not require specific mechanical properties. In the knives presented Figure 7.3.6 there is no junction between the blades and fittings, the bolsters are made out of the blade stock. Well documented and detailed information about *the knife anatomy* can be found on Jay Fisher's Internet site, a professional custom knifemaker.

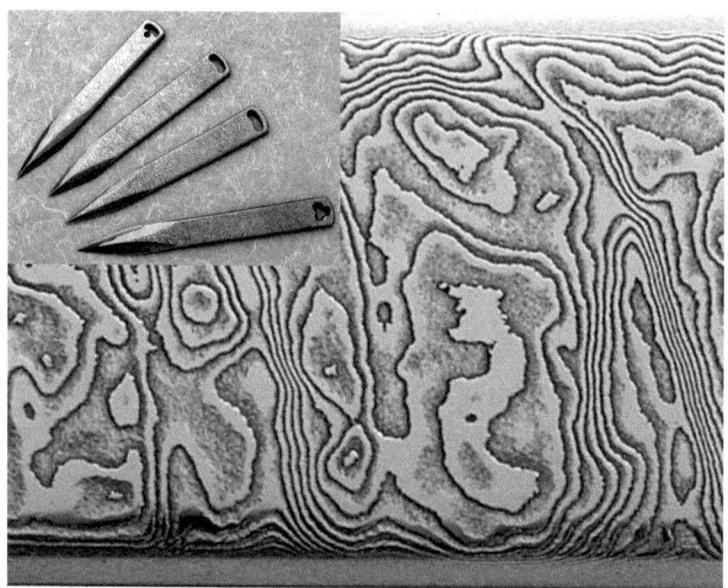

Figure 7.3.2 : Damask forged by Pierre Reverdy with three steel grades :
- core K190PM (X220CrVMo13-4) obtained by powder metallurgy ;
- wrapped in a damask made with X2CrNi18-9 (304L, S30403, 1.4307) and XC75 steel grades. Courtesy Pierre Reverdy, Romans, Fr.

The smiths use also recycled steel of diverse origins : cables, bicycle chains, cuttings etc. The twisted threads of the cables produce after currying a completely aesthetic wavelike decoration. Chains produce a decoration in which the links form a pattern. The cuttings are less easy to weld! The trick consists in locking them up in a tube in the presence of a little borax and in forging the whole.

Industrial productions on a reduced scale

Quality products were proposed in the form of thick strips of steel already damasked, intended to produce knives or various objects: alarm clocks, watches, pens etc. Knifemakers can machine blades or handles and also forge them.

The crucial step for manufacturing such pre-damask steels is welding plates of various grades, *i.e.* forming the equivalent of the stack of the blacksmith. Stainless steel plates, in particular, require a more sophisticated technique because it is necessary to take into account the

Figure 7.3.3 : Stainless steel damask forged by Friedrich Schneider. Courtesy Alain and Joris Chomillier knifemakers, Clermont Ferrand, Fr.

characteristics of each steel. There is a difficult high-temperature step in which oxidation must be avoided. Equipment for the partly industrialized manufacturing makes it possible to prepare bars of stainless grades ready to be machined or hot worked. The detailed procedures for welding these plates are kept secret by the manufacturers. Figure 7.3.3 presents a realization starting from pre-damask bars. The formation of a pattern is explained in §7.5.

Structural two-phase steels, "duplex" steels

It is necessary to mention a category of steels excellent for their mechanical and corrosion resistance : the duplex steels. They have a steep designed composition with a high proportion of several alloying elements [70]. By heat treatment at around 1000°C, they develop a two-phase structure with austenite and delta ferrite, a structure which is suspended by a specific quench in a metastable equilibrium state. By cold rolling, the structure becomes similar to that of a damask steel. Such bars could not be forged by the knifemaker, *i.e.* heated below

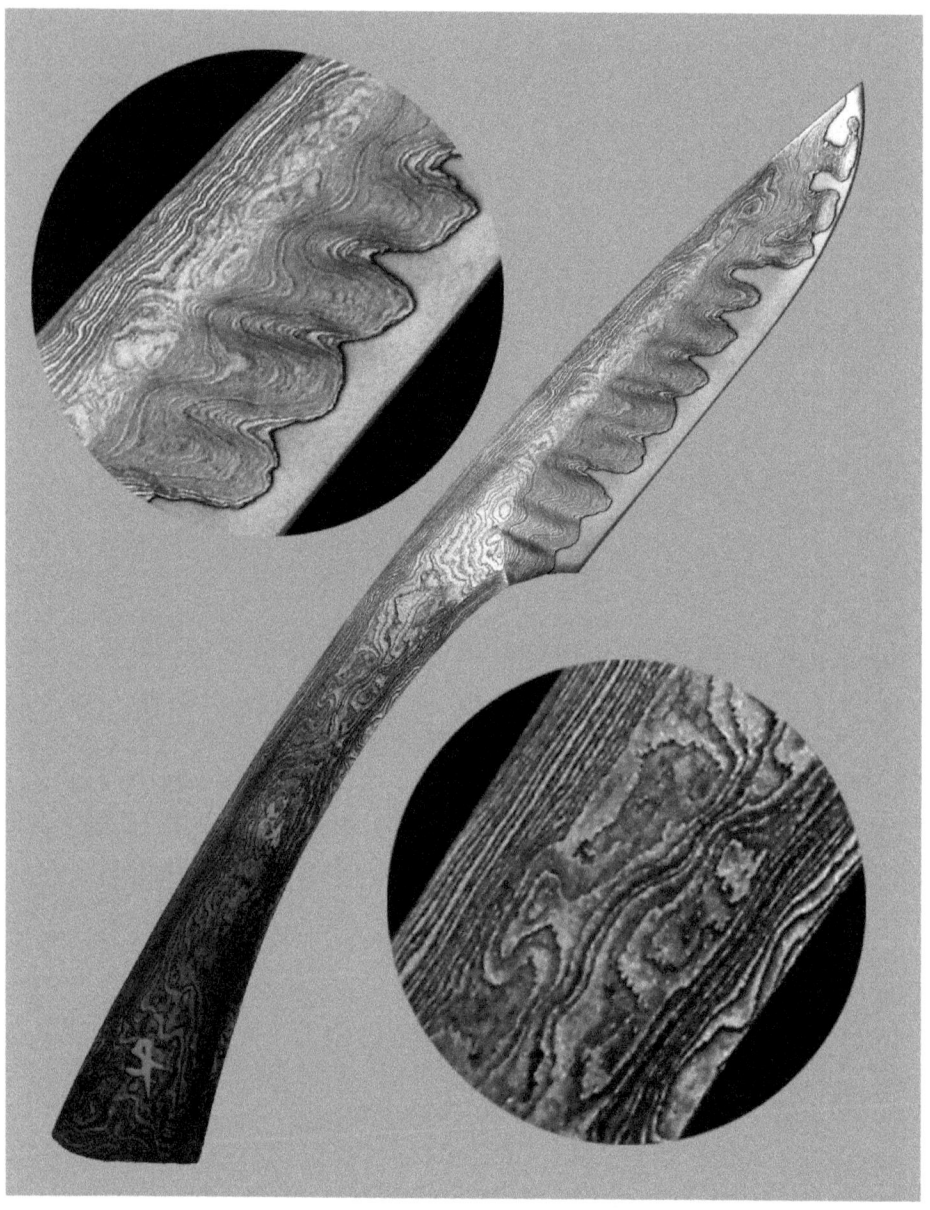

Figure 7.3.4 :
Integral damask knife forged by Matthieu Petitjean, with 90MCV8 (90MnCrV8) dark shade, and 45NCD16 (45NiCrMo16, EN 10027) bright shade with nickel.
Courtesy Matthieu Petitjean, St Clair sur Galaure, Fr.

Art and Technology in the Third Millennium

Figure 7.3.5 :
Knives forged by Jean-Pierre Veysseyre, Thiers, Fr
Courtesy *La passion des couteaux*, Fr

Figure 7.3.6 :
Knives with integral bolsters forged by Friedrich Schneider, Germany
Courtesy *La passion des couteaux*, Fr

Figure 7.3.7 : Knife made by Jerry Rados, Columbia, USA.
Notice the spine of the blade. Courtesy *La passion des couteaux*, Fr

1000°C, otherwise they lose their decorative two-phase structure. However they could be worked by cold machining followed by a surface treatment to emphasize their natural damask aspect [71] to manufacture various objects, except blades on account of a poor wear resistance.

7.4 Inserts and mosaic patterns

An alternative to the above methods consists in incorporating kinds of sculptures in a damask matrix. It is the hot co-forging of cut out blocks which produces the pattern.

The steps of the manufacturing of a decorative module are explained in Figure 7.4.1. with Pierre Reverdy's samples (his famous unicorn). Two blocks of 70x100 mm size are chosen in different steel grades: 203E (low carbon) and Fe-0.9C-2Mn-0.3Cr-0.1V.
Using EDM (Electric Discharge Machining), the design is pierced through one block of each steel. The interior patterns for each block are interchangeable, *i.e.* encasable in the matrix of the other block. After exchange of the cut out parts, the whole is welded, stretched, possibly twisted by hot forging to form a composite mini-bar of smaller cross-section. These mini-bars are then assembled in a bundle and are welded in order to form only one bar with repetition of the decoration. Figure 7.4.1 bottom presents a section of a bar made up of sixteen mini-bars. The outline of pictures of the inlays must remain very clear, then forging has to be rather rapid in order to limit interdiffusion of the elements between both steels.

Slices of these decorative modules are taken, integrated into a substrate so that the decoration appears on the flat of the blade. The unit can still be stretched or twisted so that the pattern outlines a farandole of figures as in the case of the dagger shown in Figure 7-4-2. The composite bar before being distorted is visible through the transparent veil.

Hank Knickmeyer's technique is different. Decorative bars and rods are placed inside a steel tube with a square section. Then the remaining space is filled with alloyed steel powder. The ends of the tubes are welded shut and the whole billet is weld forged in a single heating using a hydraulic forging press.

Examples of the Damascus *butterflies and flowers* mosaic pattern by Hank Knickmeyer are shown in Figure 7.4.3. The decorative mosaics are inserted on the flat part of the blade (Figure 7.4.4), while a band of twist pattern forms the cutting edge made with two grades of carbon

Figure 7.4.1 :
Top : Blocks with inserts ;
Bottom : Electron micrograph with enhanced contrast showing a set of 3cm on side comprising 16 bars welded side by side. Micrograph Grenoble INP, Fr. Samples: Pierre Reverdy, Romans, Fr.

Art and Technology in the Third Millennium 107

Figure 7-4-2: Dagger, Courtesy Pierre Reverdy, Romans, Fr

Figure 7.4.3 : Hank Knickmeyer's mosaics, Missouri, USA
Courtesy *La passion des couteaux*, Fr.

steels 1084 and L6. This choice leads to a very high quality cutting edge after the appropriate heat treatments.

Figure 7.4.4 :
Two knives forged by Hank Knickmeyer, Missouri, USA
Courtesy *La passion des couteaux*, Fr.

7.5 Sintered steels

The technique of sintering consists in welding metal powder grains (see § 8.10). It has been used for a long time to manufacture preformed steel parts, but the development to obtain multi-layer steels is more recent. Steels, named Damasteel®, were manufactured by powder metallurgy according to a process patented in 1998 [72].

The particular process to form damask steels

Two powders of different grades are placed in a capsule in stratified layers, not intermixed (Figure 7.5.1). The unit is sintered at high temperature and under an isostatic pressure of about 1000 bars. The bar or small ingot obtained is used later on as a raw material in forging.

The process offers, in theory, the possibility of combining multiple grades. For each combination a specific sintering temperature must be defined in order to weld the powder grains without melting them. Indeed, the fusion would destroy the extremely fine microstructure with a multitude of well distributed carbides resulting from the rapid solidification of the droplets (§ 8.10). It is this particular microstructure which gives the alloy its excellent properties of use.

The most difficult stage is densification because the temperature must be adapted, compatible with the two grades of steels, a temperature to which just enough liquid forms to stick the grains between them. The precise optimization of the sintering window is experimentally determined for each couple of grades.

Advantages of sintering

It is a sophisticated technology, thus expensive. However, the use of alloyed powders makes it possible to choose compositions, with much higher carbon or chromium contents, which could not be developed by the usual melting method. Chromium confers a good corrosion resistance. Carbon reinforcement operates both by martensite transformation and by precipitation in the matrix of very many small carbides on a micronic scale, mainly of $[Fe,Cr]_{23}C_6$ type (see Figure 7.5.3). These carbides inlaid on the sharp edge work like the tiny teeth of a saw. Damask steel thus obtained is technically excellent.

In addition, coarse carbides reduce the fracture strength in conventional steels. The carbide clusters act like fracture initiation sites at a certain stress level. The substantially smaller carbides in the rapidly solidified powders inhibit fracture initiation until the stress reaches a higher level.

Create a pattern

Pre-damasked steels are marketed in the form of strips intended to be cut and machined into blades by the knifemakers. In fact the pattern is created starting from a structure laminated by hot or cold work (Figure 7.5.1). Among the decorations thus obtained, the decorations of Muhammad Ladder and Rose imitate the wootz patterns (Figure 7.5.2). Skilled machinings make true sculptures which emphasize all the facets of the decoration.

The revelation of the microstructure is difficult because stainless steels are by definition resistant to chemical attack. The etchants are necessarily strong acids: hydrochloric, perchloric, nitric or sulfuric. Mixtures can be dangerous (corrosives, explosives) also it is essential to follow the recommendations of the manufacturer [73].

Forging is possible respecting the limitations of temperature indicated by the manufacturer. In the case of the blade shown on Figure 7.5.3, the manufacturer reports beginning of fusion at 1220°C. An excessive heating could ruin the alloy. Consequently a good control of the heating temperature is needed; electric or gas fired furnaces are recommended. Two grades used in the production of Damasteel® are: RWL 34 and PCM 27 which have the compositions in wt% :
(RWL 34) 1.05C -14Cr -0.5Mn -4Mo -0.5Si -0.2V
(PMC 27) 0.6C -13.5Cr -0.5Mn -0.5Si .

RWL 34 is closer to a tool steel than to a steel for knife blades since it is rich in carbon, chromium and molybdenum. It is the harder steel of the couple and finishes bright after etching, while PMC 27 is softer and dark after etching. Global hardness may reach 60 HRC.

Blacksmiths integrated these new alloys in traditional creations associating precious materials (Figure 7.6.2 with a Damasteel® blade to be compared with Figures 7.6.1 and 7.6.3 with traditional damasked blades).

1- filling
2- encapsulating 3- hot isostatic pressing 4- forging 5- rolling

Figure 7.5.1 :
Steps for making the bar :
1- layers of two metallic powders of different grades are placed in a metallic capsule ;
2- the capsule is sealed ;
3- the unit is sintered at high temperature and under isostatic pressure of 1000 bars (see image of two bars);
4- the bar is hot forged ;
5- the bar is hot rolled.

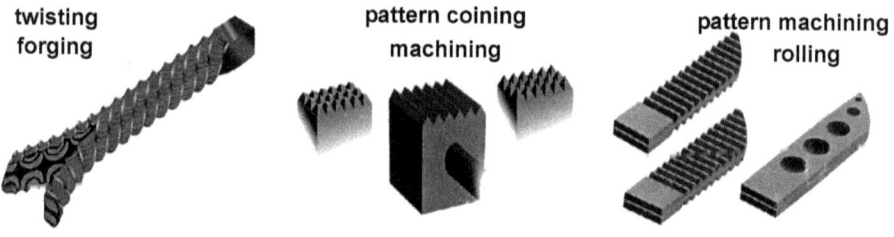

twisting forging   pattern coining machining   pattern machining rolling

Figure 7.5.1 :
Formation of a pattern.
- hot twisting and stretching;
- cold stamping and machining. This method makes it possible to create the decorations in waves and undulations. It consists in deforming the layers by stamping, then levelling the surface.
- cold machining and hot rolling. This method consists in removing matter by machining. Hot rolling levels the surface. It is similar to that used for the wootz in which the blows or the extra thickness are intended to disturb the regularity of the layers to create the decoration with ladders or roses.
Courtesy Damasteel®

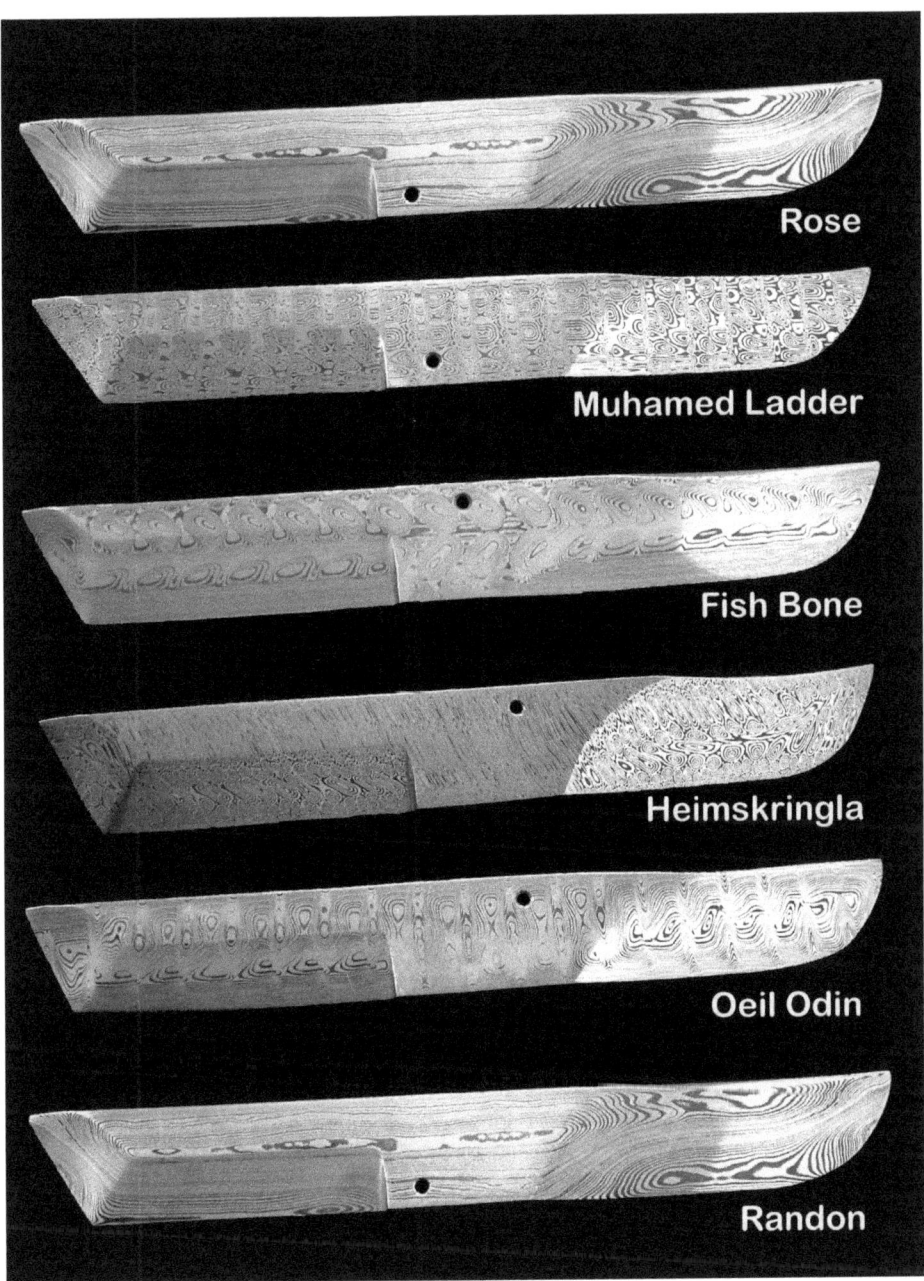

Figure 7.5.2 :
Samples of the presentation of the Damasteel® decorations. Each bar is cut according to various planes making it possible to see all the variants of a decoration. Adapted after Eurotechni documents, La Monnerie, Fr.

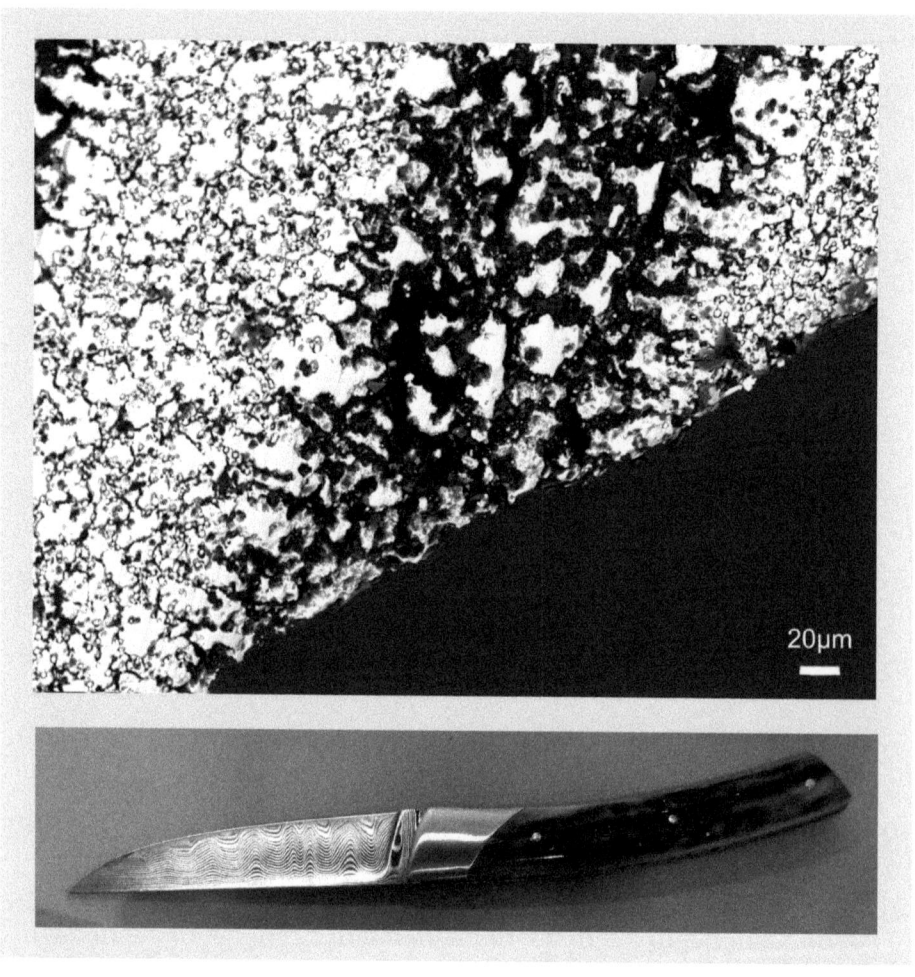

Figure 7.5.3 :
Knife with a Damasteel® blade. The blade was etched to reveal the pattern. Electron micrograph shows fine precipitation.
Document Grenoble INP, Fr. Knife Chambriard, Thiers, Fr.

7.6 The colored blades

Both iron and steels oxidize. Even steels said to be stainless can be oxidized under severe conditions (see §8.9). The first stages of oxidation form very thin layers appearing with a blue-purple color which evolves to the brown russet-red depending on the thickness. To create a colored decoration it is necessary to control perfectly the kinetics of the attack of the surface which depends on the operating conditions and the steel composition. Thus, in a damask steel, the thickening of the layers of each constitutive steel follows its own progression of color. The pattern is emphasized by an intense contrast of colors. The chemical attack of steel is ready in the case of non-stainless steels, but it is also more difficult to control precisely. The samples must be polished carefully because the smallest scratches are highlighted by colorings. The fine layer of oxides is hard and quite resistant but it can be damaged by abrasives. The knife in Figure 7.6.1 is made by the traditional method of pattern welding, starting from two steels with different carbon contents. The sophisticated coloring is the result of electrochemical treatment. In the case of the knife in Figure 7.6.3 a high contrast of color is rather easy to obtain on account of the high corrosion resistance of nickel.

Figure 7.6.1 : Damask steel knife. Courtesy François Pitaud, Les forges Pitaud, Fr.

Art and Technology in the Third Millennium

Figure 7.6.2 :
Knife with brown handle in horn of mammoth and a Damasteel® blade. Blueing is obtained by heat treatment at 510 °C. The wide bolster was machined out of meteoritic iron (originated from Gibbeon).
Courtesy Des Horn, Onrurivier, South Africa.

Figure 7.6.3 :
Knife with handle out of mother-of-pearl, classical damask blade and wide bolster made with nickel and steel damask 15N20 Sandvik. Blueing is obtained by heat treatment at 290 °C.
Courtesy Des Horn, Onrurivier, South Africa.

7.7 The contemporary wootz

Reconstituted wootz

Reconstruction of the structure of ancient wootz has been performed successfully by many researchers, in particular by the Verhoeven-Pendray team [59]. Image 7.7.1 is representative of this experiment which evidenced that the moire pattern is formed by cementite precipitates of a few micrometers distributed in regularly spaced rows on account of residual segregation of crucible steel.

Some contemporary blacksmiths, of which Achim Wirtz (Germany) is one, took up the challenge to manufacture beautiful blades starting from crucible steel. The image of Figure 7.7.2 shows a blade forged in reconstituted wootz, *i.e.* in as-cast steel similar to the ancient crucible steels with iron and carbon and without an alloying element. The preparation was carried out by fusion of iron in a graphite crucible (Figure 7.7.3). There is a reaction between iron and graphite and a certain proportion of carbon is dissolved in the liquid iron. Another reconstruction of ancient wootz was performed by Greg Obach (Canada). The different steps of steel preparation in a graphite crucible, and the detailed hammer forging of the blade to obtain a *Damascus Type* blade are published on the Internet [74-75].

The moire pattern is more difficult to obtain with alloyed steels than with carbon steels under the same conditions of craft forging. The blacksmiths claim that a significantly larger number of operations of heating and forging without folding is necessary, thirty five operations in the case of the example in Figure 7.7.4.

This is the end of the description of the different techniques developed to produce artistic blades. Many aspects must be discussed on a metallurgical point of view, in particular the formation of carbide alignments, a very controversial subject during the last thirty years. Sound explanations can be now proposed, however the argumentation involves complex metallurgical mechanisms. Consequently this discussion is reported in chapters 9, 10 and 11 after having revisited the fundamentals necessary for the discussion in chapter 8.

Figure 7.7.1 :
Dagger forged by Al Pendray. USA. [59]
Microstructure of the blade showing alignements of carbides in dark contrast.
Document of Iowa University, USA.

Art and Technology in the Third Millennium

Figure 7.7.2 : Wootz blade with 1.9 %C steel and detail of the blade.
Courtesy Achim Wirtz, Würselen, Germany.

Figure 7.7.3 :
Small crucible cast steel ingot in a graphite crucible (inserted image). The surface of the ingot, once cleaned, observe the network of dendritic branches. Documents of Achim Wirtz starting from his own experiences

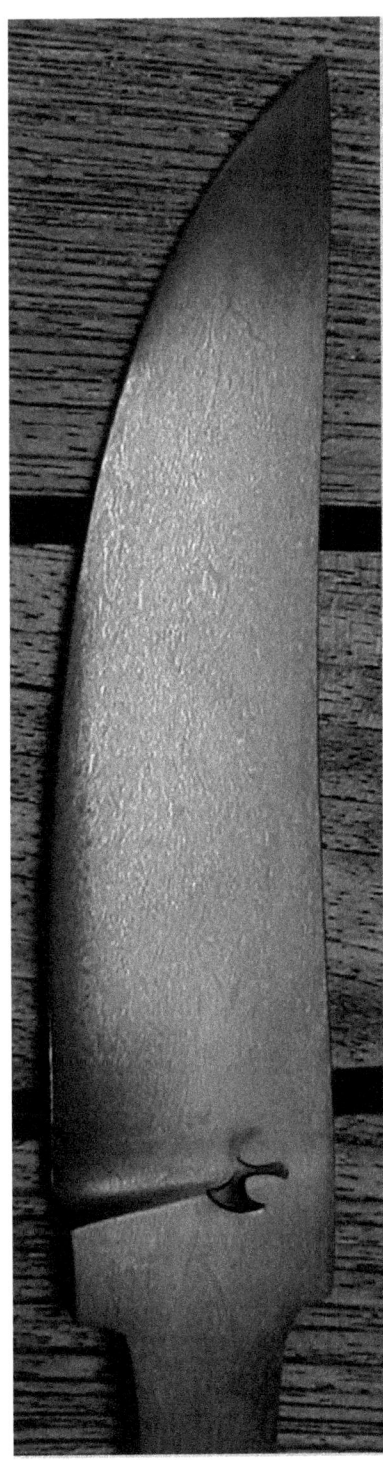
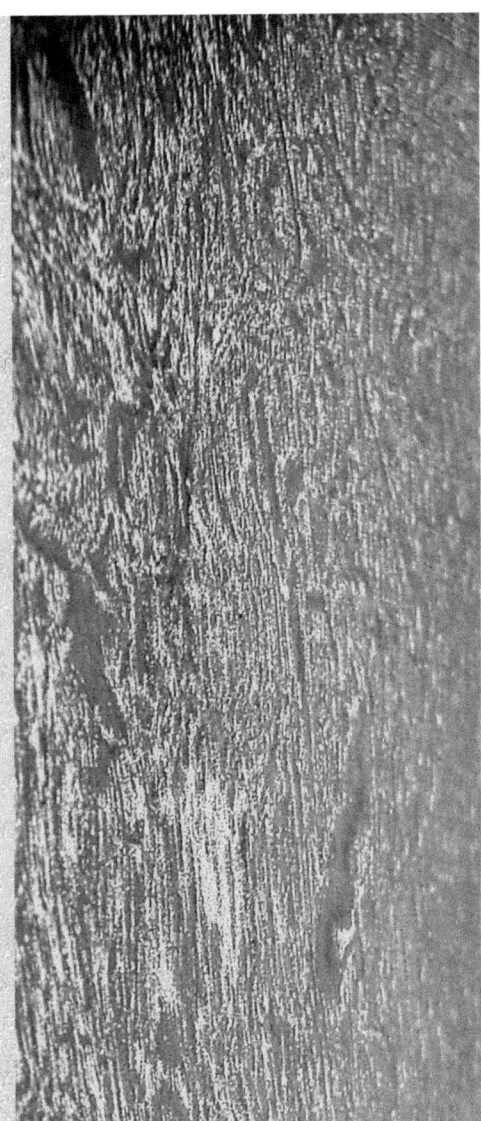

Figure 7.7.4 :
Wootz blade with 1.8 %C, 15%Cr steel. Thirty five operations of heating then forging are necessary to obtain a visible pattern.
Courtesy Achim Wirtz, Würselen, Germany.

Second part: Formation of the damask pattern

The second part is a discussion of the mechanisms involved in the damask pattern formation. Three main categories of microstructures are distinguished: the laminated structures obtained by forge welding, the moire structure of carbide alignments obtained with high carbon steels and another wavy pattern obtained with medium and low carbon steels. It appeared that it was necessary to define a metallurgical language to be understood by readers having different approaches to the subject. Chapter 8 presents a brief introduction to the basic concepts useful to support the discussion. In addition, as the most controversial points concern high carbon steels which undergo complex transformations, a particular emphasis is given to the expertise of their specific microstructural features. Other metallurgy textbooks can be read to supplement the fundamentals of metallurgy: [5, 70, 76-78].

8 *Understanding steels*

8.1 Phases and phase diagrams

Phase, solid solutions and interstitial or substitutional solutes

An alloy can be formed of one or several phases. A phase is characterized by a stable and homogeneous structure in a given interval of temperature and in a given interval of composition. These intervals define the field of the stability of that phase.
For example, in the range of temperature -10 to 110 °C pure water may appear under three different phases at atmospheric pressure :
- solid between -10 and 0 °C,
- liquid between 0 and 100 °C,
- vapor between 100 and 110 °C.
Pure water and salt water are the same liquid phase whose field of stability extends until the limit of saturation in salt defined for each temperature.

The solid phases are characterized by their crystallographic description, *i.e.* the geometry of the repetitive pattern for the crystal, the nature, and the position of atoms. The composition is not a determining characteristic because it often corresponds to several possibilities of crystalline arrangement according to the range of temperature. Pure iron has a face-centered cubic crystal structure between 912 and 1394 °C. This phase is called γ-Fe, named *austenite* after the eminent English metallurgist W.C. Roberts- Austen). Pure iron has a body- centered cubic structure called *ferrite* both between two ranges of temperatures, a high temperature δ-ferrite between 1538 and 1394 °C and a low temperature α-ferrite below 912 °C.

In steels consisting in iron and carbon, it is necessary to consider the phases called carbides which have an important hardening role. The main carbide in unalloyed or low alloyed steels is cementite Fe_3C.

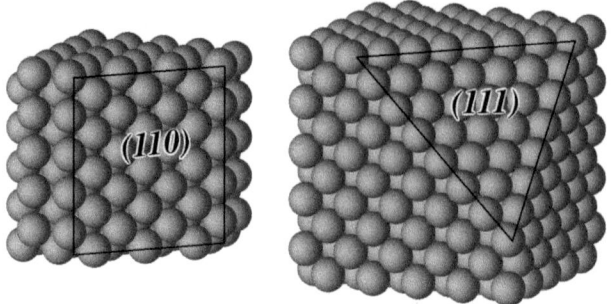

Figure 8.1.1 :)
Atomic packing in ferrite (left) and in austenite (right) structures. The sectioned planes are those of densest packing, *i.e.* in ferrite the rows of atoms draw rectangles (twofold symmetry) and in austenite the rows draw equilateral triangles (threefold symmetry).

When the chromium content is higher than 12 % in the case of stainless steels, carbides are complex compounds, $(Fe,Cr)_{23}C_6$ and $(Fe,Cr)_7C_3$.

A phase is defined by its crystal lattice, different atoms may fill specific sites. These atoms, other than iron (alloying elements, impurities), form a solid solution, they are called solutes. From the structural point of view, it is usual to consider the iron crystal as a stacking of hard incompressible spheres, each atom having a specific number of neighbors. In Figure 8.1.1 each atom is in contact with 8 neighbors in the ferrite structure, and with 12 in the austenite structure. Between the atoms there are interstices liable to lodge atoms smaller than iron, such as hydrogen, oxygen, boron and nitrogen. Atoms which occupy interstices are said to be *interstitial*. These sites, with octahedral or tetrahedral shapes, are more or less spacious according to the arrangement of their neighbors in the crystal stacking. Austenite comprises many interstices able to comfortably welcome carbon atoms up to a significant number. Ferrite comprises a higher proportion of interstices which are smaller and consequently accepts only a very limited number of carbon atoms.

The bulky atoms cannot be accepted into interstitial hollows, they can just be substituted for an iron on a normal crystal site, they are *substitutional* atoms. The phase is called a substitutional solid solution. The presence of atoms different from iron changes the mechanical properties of the solid solution significantly. The effect depends upon the nature of the foreign element and whether the arrangement is randomly distributed or regularly ordered involving particular bonding in the crystal lattice. Mostly, they are added with the purpose of reinforcing the solid solution as molybdenum, for instance, in austenite.

Diffusion

In a crystalline solid, atoms move either by jumping into empty adjacent sites (a normal vacant site or an interstitial site), or by changing place with normal sites unoccupied. Small size atoms are the only ones able to migrate through the interstitial sites. The larger atoms are forced to migrate to the few vacant sites, called vacancies, in turn releasing a new site. The frequency of jump strongly depends upon the temperature for the same diffusing element. In the case of high temperatures it can reach colossal figures, about several billion times a second.

However the atomic movements are randomly made in all the directions, so that on a macroscopic scale the average displacement is much more limited. It is this apparent shift that determines the penetration of an element in the lattice. The small interstitial atoms migrate easily, their distribution is spread out with time. Consequently, in the case of carbon, a homogeneous distribution of the carbon can be rather quickly obtained in temperatures above 900 °C approximately.

On the other hand, in the case of substitutional solute elements, the inhomogeneity of distribution or segregation remains much longer. It takes much higher temperatures and longer times to erase it. In Table 8.1.2, the diffusion ability of carbon and some substitutional elements in iron are compared. Around 910 °C, iron can be stable either under a ferrite structure, or under austenite structure. The depth of penetration of carbon is one thousand times larger than that of the other elements and it is five times larger in ferrite than in austenite.

Table 8.1.2 : Depth of penetration after a one hour hold around 910 °C, after [70].

Element	Depth of penetration (µm)		Comment
	within ferrite	within austenite	
C	1000	200	distance of the same order as a grain size
Mn, Ni, Cr	3-6	0.2-0.4	

Phase equilibria

In real material samples, the presence of particles of different phases may be discussed by reference to phase diagrams. At a given temperature T, a sample of a given composition c has a **unique** equilibrium state, which is represented on the phase diagram (T, c). When the kinetics of transformation is too fast, as in the case of solidification for example, the equilibrium state does not have enough time to settle at a long distance with respect to the atomic scale and it is established only at a short distance ahead of the transformation front. Sometimes a temporary equilibrium settles down, the state is said to be *metastable* and finally the system tends to return to the equilibrium state when heated again.

The basic system for steels includes iron and carbon. The role of carbon in forming steel has been understood only since the middle of the 19th century. The first outlines of the Fe-C phase diagram date back to 1895–1899. This system has two basic variants, one said to be stable in which carbon is in the form of graphite and the other called metastable in which carbon is in the form of cementite Fe_3C carbide (see Figure 8.1.3). The iron-graphite system corresponds to alloys based on grey cast irons. Graphite is formed when the cooling rate is very low or after holding at high temperature for a long time. Its formation is induced by the presence of certain elements said to be graphitizing such as silicon, nickel and aluminum.

The Fe-C diagram is often presented with a series of triple lines. Two lines shifted above and below the lines representing the equilibrium correspond respectively to the temperatures of transformation on heating and cooling, they recall in a purely qualitative way that each transformation starts with a certain delay.

In the case of the binary system (two constituents) Fe-C three main reactions occur on cooling at only one temperature involving three phases (see Table in Figure 8.1.3). One of the phases disappears at the end of the reaction :
- *peritectic* reaction P3, δ ferrite (F3) disappears;
- *eutectic* reaction E2, liquid (L2) disappears;
- *eutectoid* reaction E1, austenite (A1) disappears.

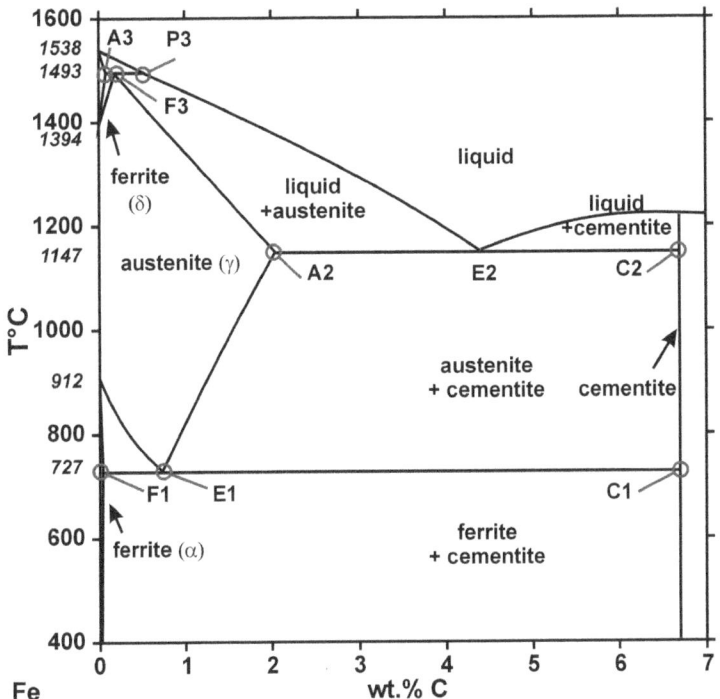

Figure 8.1.3 : The Fe-Fe$_3$C system
This diagram represents the conditions (Temperature T, Composition c) for which the different phases can be in equilibrium. For example austenite γ can be in equilibrium as single phase only for (T, c) conditions represented in the field on the left side of the diagram. Single phase fields exist also for δ-ferrite, α-ferrite, cementite and liquid. In the case of α-ferrite the phase field is very narrow and represented by a thick line. Cementite is represented by a line since the formula is strictly restricted to the 1/3 ratio.
In two phase fields there exist two phases, the proportion of which is determined for given (T, c) (known as inverse segment rule [70])
Three phase equilibria are represented by a point, meaning that composition and temperature are unique for these reactions which are indicated E1 for eutectic, E2 for eutectoid and P for peritectic.

Reactions	Phase compositions for invariant reactions (wt % carbon)	T°C
Eutectoid E1	F1 ferrite (0.022)-E1 austenite (0.76)-C1 cementite (6,69)	727
Eutectic E2	A2 austenite (2.14)-E2 liquid (4.3)-C2 cementite (6,69)	1147
Peritectic P3	F3 ferrite (0.09)-A3 austénite (0.16)-E3 liquid (0,53)	1493

Calculated diagram, Courtesy Grenoble INP, Fr

The eutectic reaction which forms austenite and cementite in a coupled way develops a specific ladderlike morphology (ledeburite) with a very hard and rigid structure whose lack of ductility makes it difficult to forge. This is why the carbon content is limited in ancient and modern steels to about 1.5%C.

8.2 Austenite transformation in the Fe-C system

The austenite is the stable phase in a large domain at high temperature, it is an interstitial solid solution of carbon in face centered cubic iron. At temperatures under that of the eutectoid transformation it decomposes into different constituents according to the cooling rate.

1- A very low cooling rate enables the transformation into the stable phases, which are ferrite and graphite.

2- With a moderate cooling rate austenite transforms into ferrite and cementite. As these are not the most stable phases the equilibrium is said to be metastable.

3- A high cooling rate, a quench for instance, transforms austenite into martensite, a metastable phase with the same composition, but a different crystal structure. The transformation consists in a change of the atomic arrangement without carbon diffusion.

Pearlite

The eutectoid reaction at 727°C results in the coupled cementite and ferrite growth. The product of this reaction is called pearlite. The pearlitic microstructure is made of cells in which cementite and ferrite plates alternate (see Figure 8.2.1). The thickness of the plates can vary between a few tens of nanometers and some micrometers. This structure is relatively ductile. Several examples will be presented next. The presence of other elements in addition to carbon as manganese can facilitate the formation of the pearlite, but conversely many others can delay it by inhibiting nucleation or growth. Silicon and nickel have such an effect even with low content.

The formation of pearlite is a very exothermic reaction. The blacksmiths detect it easily by observing that the piece of iron they are forging, changes color during cooling.

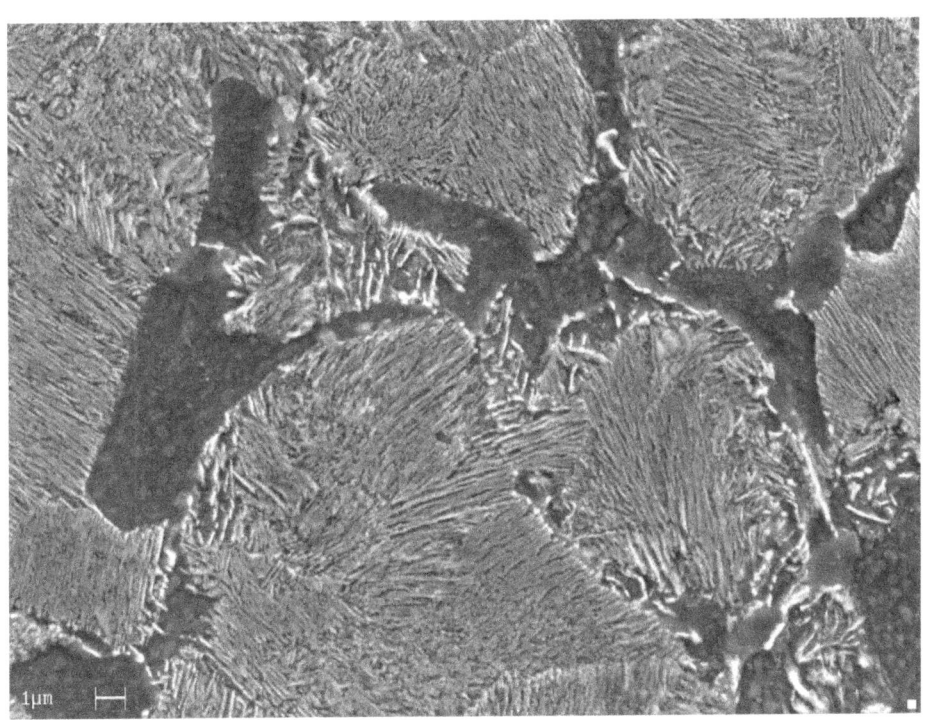

Figure 8.2.1 :
Electron micrograph illustrating very fine pearlitic cells.
Courtesy Grenoble INP, Fr.

Martensite

Martensite has the same composition, therefore the same percentage of carbon as the original austenite. Its quadratic centered crystallographic structure can be considered as a supersaturated ferrite which is distorted by the presence of carbon in the interstices. This phase is very hard. To give an idea of the scale of hardnesses Table 8.2.2 presents some values of microhardness.

Table 8.2.2 : Vickers values for microhardness

Phase	Diamond	SiC	Fe_3C	Martensite	Austenite	Ferrite
microhardness H_V	8000-6000	3500	1200	500-1000	190-350	70-190

Martensite grows very quickly, with a speed about the speed of sound. The formed plates or laths are elongated in certain preferred directions for which the growth is easier taking into account the crystallography of the matrix. Ordinary steels develop martensite with sheaves of laths illustrated (Figure 8.2.3 Top). High alloyed or carbon rich steels develop a martensite in individual plates which leave areas of untransformed austenite between them (Figure 8.2.3 Bottom).

Martensite can be formed only below a characteristic temperature called Ms (meaning martensite start) and the transformation is completed at temperature Mf (martensite finish), some 215°C below Ms. If the alloy is brought down to a temperature between Ms and Mf the transformation is incomplete and there remains some residual austenite. The austenite/martensite mixture is not desirable because this heterogeneous association is prone to rupture. It must be avoided when good mechanical performances are required. However, carbon and almost all alloying elements lower the values of Ms and Mf and thus facilitate the austenite retention.

The choice of a steel with a moderate percentage of carbon, lower than 0.6%, ensures a temperature (Ms) higher than 250°C, therefore an easy quench hardening. But in the case of a 1% carbon steel (see Figure 8.4.2), Ms is below 200°C. Consequently a quenching process at room temperature will leave the transformation incomplete.

The effect of the quenching temperature was indicated in 1835 in Landrin's treaty [52] : *All the smiths noticed that steel was much harder in the cold and during the frost than during the warm season.*

Nowadays, there are two answers to obtaining the complete transformation of very high carbon steels : either several repetitive operations of quenching and tempering (see § 8.8) or quenching at a temperature lower than Mf, in liquid nitrogen, which acts only as a simple cooling fluid instead of air, oil or water.

Bainite and other constituents

Bainite is a structural constituent formed of carbides and ferrite enriched in carbon. Its hardness ranges between 300 and 500 Hv *i.e.* intermediate between the hardness of martensite and that of pearlite. Polemics between scientists have been regularly reactivated for almost 80

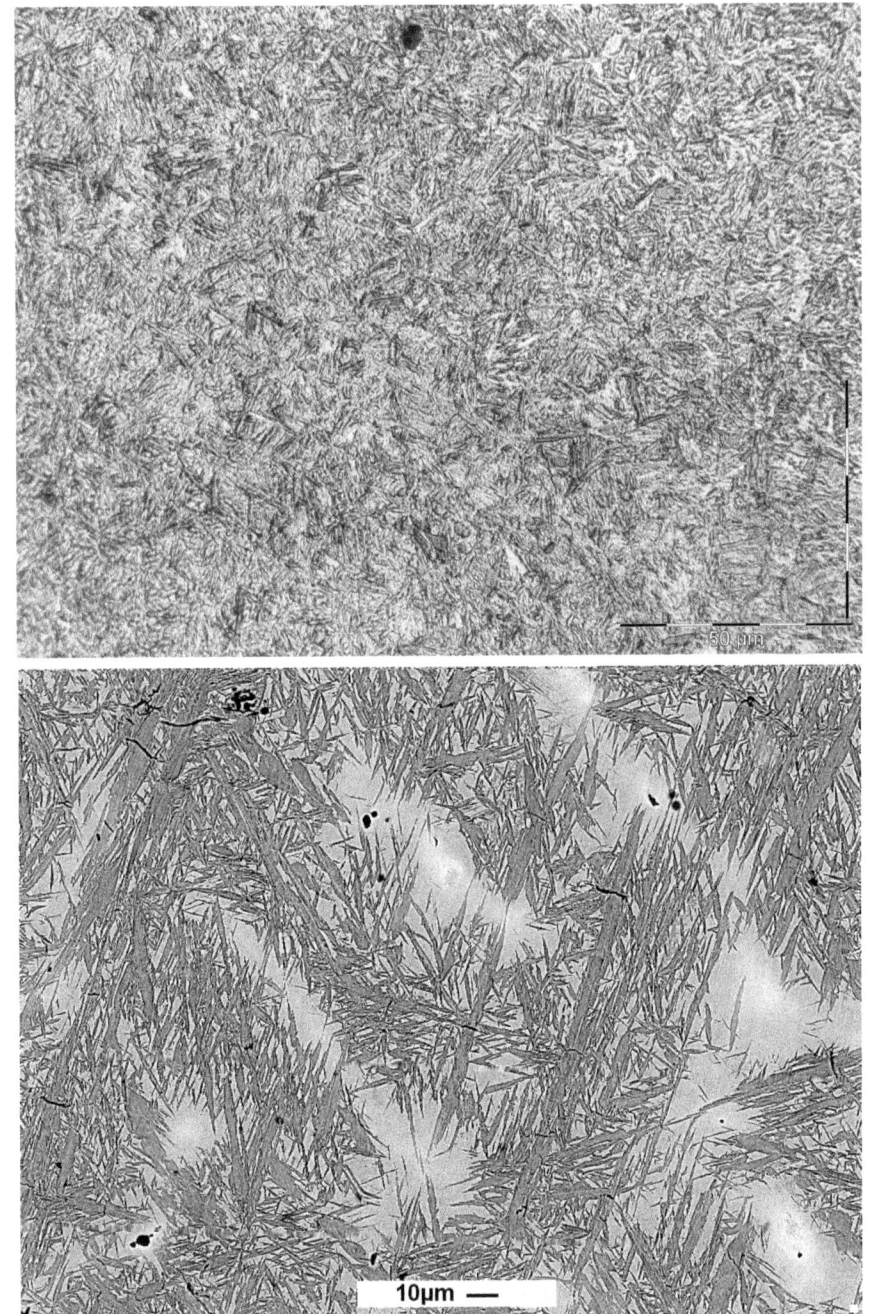

Figure 8.2.3 : Electron micrographs
Top : lath martensite; Bottom : plate martensite (high magnification). Courtesy Grenoble INP, Fr.

years concerning the mode of formation of bainite [70]. To simplify, let us admit that the microstructure of upper bainite is close to that of pearlite and that of lower bainite is close to that of martensite. The name of troostite, previously employed, corresponded to a poorly defined mixture of various components.

8.3 Kinetics of the austenite transformation

Continuous cooling transformation (CCT diagrams)

The presence of alloying elements plays an indirect role on the martensite formation. In fact, it is the kinetics of the pearlitic transformation which is changed, in most cases delayed. The most practical representations are the continuous cooling transformation (CCT) diagrams. The couple of points representing the transformation start into pearlite, bainite or martensite and the corresponding transformation finish are plotted in temperature-time diagrams along curves representing different cooling profiles. These diagrams must be read exclusively along the cooling curves. They are plotted on a logarithmic scale (Figure 8.3.1). The starting and finishing points determine a field in which a specific transformation progresses until it stops at the exit of the field with a certain rate of completion noted near the curve. Thus for the curves in dotted lines which correspond to the slowest cooling rates, austenite is completely transformed into pearlite well before reaching the Ms temperature. The exit of the field of the martensitic transformation Mf is not indicated because it is below 0 °C for high carbon steels.

Practical information on CCT curves is the mention of final hardness for each cooling rate. A high hardness is obtained in the case of fast cooling rates according to the curves with solid lines whereas it falls considerably for longer times according to the curves with dotted line.

The curve in Figure 8.3.1 A is seldom presented because it corresponds to a steel without element of addition, that makes it very difficult to transform into martensite. The transformation into pearlite is practically immediate thus unavoidable. The curve is exceptionally spread over the logarithmic scale of time. The interest of this curve is that it makes clear the low hardenability of ancient steels like those of the Celts prepared by smelting of iron ore.

Figure 8.3.1 : A
Simplified curve of continuous cooling transformation CCT Fe1%C steel, adapted from the *Atlas zur Wärmebehandlung der Stalhe*.
Samples were austenized during 30 minutes at 850°C.
Coolings follow the grey full curves and the dotted curves. The small figures close to a line of transformation indicate the percentage of austenite transformed in the previous field.
Final hardnesses after quench and 2 minute hold are reported in the lower bar either in HRC Rockwell units, or in Hv Vickers units (Hv in this case and usually for the low values).

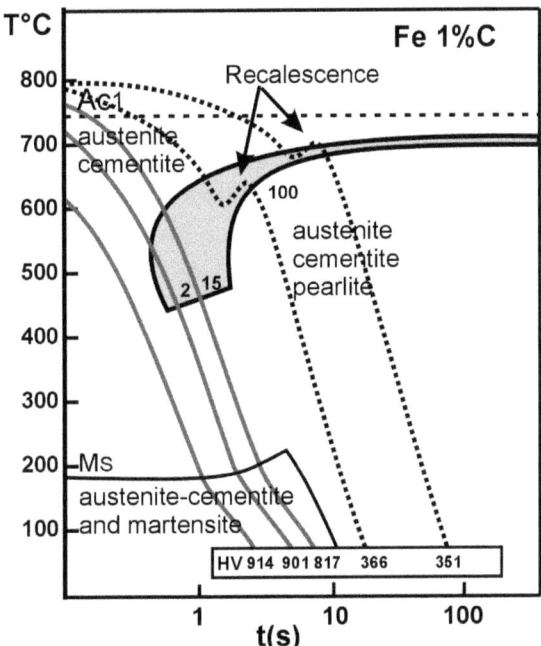

Figure 8.3.1 : B
Simplified curve of continuous cooling transformation CCT of 100Cr6 steel (1C-0.3Mn-1.71Cr-0.04Mo-0.14Cu), adapted from the *Atlas des courbes de transformation des aciers de fabrication française, IRSID*.
Samples were austenized during 30 minutes at 850°C.
Final hardness is reported in the lower bar for each cooling rate either in HRC Rockwell units, or in Hv Vickers units for the lowest values (Hv228<HRC22).

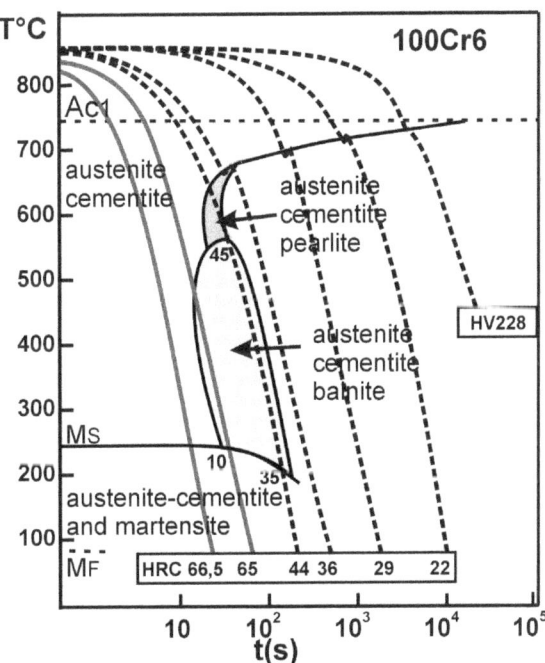

It is interesting to compare this unalloyed steel with a classic modern steel, steel 100Cr6 labeled also L3, 52100, A573Gr70, which has the same carbon content (Figure 8.3.1B, note that timescales are different). The low content of alloying elements in 100Cr6 is sufficient to delay the pearlitic transformation and consequently facilitate the martensite formation. In fact, modern steels contain specific additional elements (molybdenum, nickel) or residual impurities specific to their manufacturing (silicon) which make the martensite formation easier.

Austenization temperature

Austenization is a thermal treatment, it consists in heating the steel in the temperature range in which austenite forms. The chosen temperature imposes a physico-chemical equilibrium between austenite and carbides, indeed a definite composition of the matrix. To take again the example of steel 100Cr6, the higher the temperature of austenitization, the lower the proportion of undissolved carbides. In addition, carbon and other elements in solution in the matrix delay significantly solid state transformations. Figure 8.3.2 illustrates the modifications for temperatures of 850 and 1000 °C : the time shift to form the same components reaches almost a factor of ten. The choice of the austenitization temperature has many consequences which are analyzed further in §8.8 about stainless steels.

The cooling curves plotted on CCT diagrams are established considering cooling is carried out in contact with a given fluid, water for the previous examples, air, or heated oil. Three characteristics are to be considered for the choice of a quenching fluid (sometimes named quenchant) :
- range of temperature in which it is usable,
- its thermal conductivity. The thermal conductivity of pure water is lower than that of salted water, the conductivity of oil is low.
- its aptitude to form a thermal barrier between the object and the fluid. Heat transfers are slowed down significantly when the fluids evaporate in contact with the hot part during quenching in water or liquid nitrogen. Stirring the fluid makes it possible to limit this effect. Lastly, the quenching temperature determines the rate of completion of the transformation (Mf) and consequently the induced mechanical stresses.

Figure 8.3.2 :
Superposition of two CCT curves for steel 100Cr6 for different austenization temperatures of 850 and 1000°C during 30 minutes.
After austenization at 1000°C all the carbides were dissolved, the composition of the matrix is identical to the nominal composition.

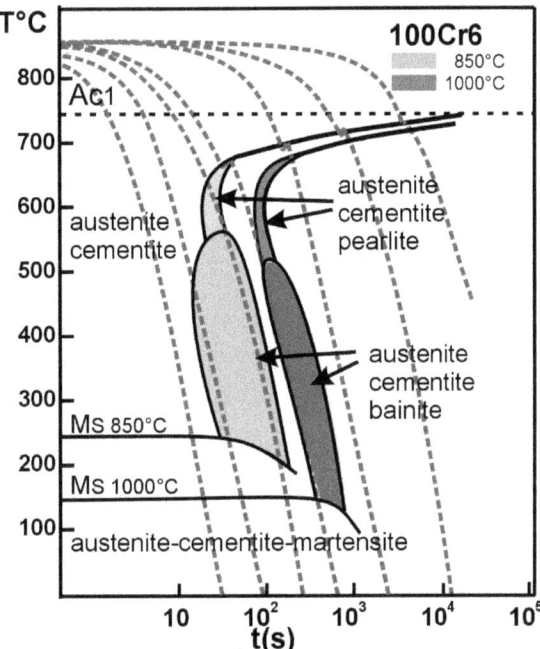

When a blade whose edge has been covered with a coating is quenched, it has a tendency to curve into a saber shape. This is due to the martensite transformation which causes a significant expansion in the case of high carbon steels. Sometimes an unexpected reverse curvature occurs, the phenomenon is difficult to explain taking into account the number of parameters.

However, let us point out a parameter perhaps neglected in traditional smithing practice, it is the austenization time. The CCT curves are established under specific conditions, in particular after a temperature hold sufficient to ensure thermodynamical equilibrium. In the case of samples protected by a coating layer, the heat transfer rate is decreased both on heating and cooling processes, then secondary carbides are not dissolved and the equilibrium is not reached in the thick part which consequently transforms more readily into martensite.

8.4 Heat treatments

Precipitation in martensite

Isothermal treatment of metastable martensite tends to restore ferrite and carbides with a specific structure depending upon the temperature range of the treatment. Heating between 100-200°C, which is a relatively low temperature induces short distance carbon diffusion and a nanometric scale precipitation. Above 100°C the carbon atoms leave the interstitial sites and move in different ways, tending to segregate in clusters along certain crystallographic planes. Around 200°C these carbon atoms form various carbide precipitates with carbon atoms in a crystallographic structure different from that of cementite. These precipitates measure some tens of nanometers and they have a rough formula $Fe_{2.5}C$. They nucleate very easily because they build together in continuity with the crystal lattice of martensite (they are said to be coherent). Simultaneously carbon depleted martensite becomes less deformed by the occupancy of the interstitial sites and as a consequence less brittle. The formation of carbon clusters or coherent precipitates is accompanied by a slight drop in hardness, but the essential characteristics of martensite are retained.

Between 250-350°C the Fe_3C cementite can be formed but, owing to the fact that the diffusion operates at a short distance, the precipitates which are formed tend to nucleate between the martensite laths and to adopt a rodlike morphology. The resulting microstructure is brittle and the phenomenon is often called *500°F embrittlement*. The structure is weakened. This temperature range must be avoided for heat treatments.

Between 400-650°C cementite precipitates within martensite laths depleting its excess of carbon. Then the precipitates coarsen and spheroidize and martensite returns progressively to the state of ferrite.

Softening and stress relieving near 200 °C

The total transformation of austenite into martensite corresponds globally to an expansion which generates internal stresses. The appropriate treatment consists in holding temperature at around 220°C for approximately an hour. The treatment must be precise in duration and temperature (±20°C) because a lower temperature is not sufficient to form precipitates of $Fe_{2.5}C$ and a higher temperature forms too large precipitates, less hardening because they lose their coherency with the matrix. This treatment improves the steel toughness. However, the temperatures and durations of the treatment depend on the grade, this is why it is recommended to follow instructions adapted to that particular steel. The best mechanical properties require optimizing the size of precipitates.

Tempering

The purpose of this treatment is to make steel very ductile and easily machinable. When a steel is heated near 500-600 °C martensite loses carbon and transforms into ferrite in approximately one hour while carbon forms with the iron of coarse cementite precipitates which do not induce any hardening. The martensitic characteristics disappear gradually according to the temperature and to the duration of the treatment. Paradoxically the structure is no more martensitic in a crystallographic point of view, but the trace of martensitic laths or needles can be observed. This morphology is sometimes called sorbite, the phases formed are the same as the ones in pearlite with a different size distribution (Figure 8.4.1). However pearlite spheroidized during a low cooling rate, may exhibit a structure rather similar to that of tempered martensite though much coarser.

Refinement of martensite

Certain treatments aim to create a finer microstructure. One operation (sometimes called standardization) consists in heating steel a few degrees above A1 (eutectoid temperature at 727 °C) for a short time. Steel is transformed into austenite without redissolving all the phases, in particular carbides. During further cooling, these phases present in the matrix constitute at the same time sites for nucleation and obstacles for the formation of large martensite laths or coarse pearlite cells. It results in a finer microstructure, globally well distributed, more isotropic, consequently having improved mechanical properties.

Specific treatments in the case of high carbon steels

The transformation of a carbon-rich austenite into martensite corresponds to an expansion which is the stronger as the content of carbon dissolved in austenite is high. This expansion can reach 5 volume%. The internal stresses thus generated can be colossal. As-cast

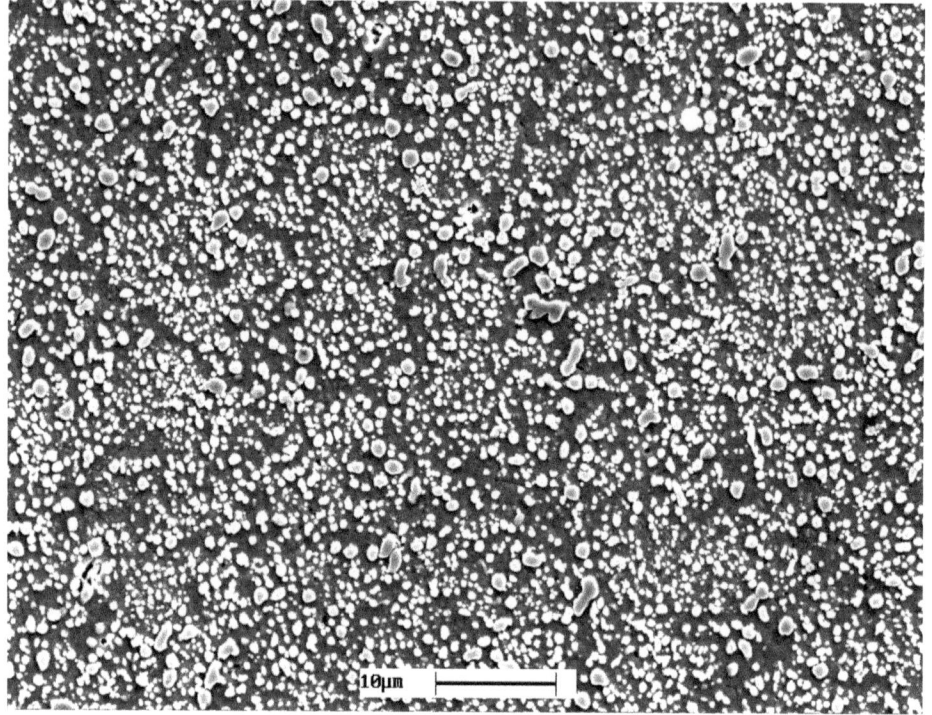

Figure 8.4.1 :
Structure after tempering. Electron micrograph of an etched sample showing carbides in clear contrast.
Courtesy Grenoble INP, Fr.

structures can be extremely brittle. Large high carbon steel castings such as rolling mill rolls present the risk of explosive stress relieving. In the case of a knife blade, quenching can cause the breaking of the blade.

The transformation into martensite remains often uncompleted on account of the stress induced by the expansion. The untransformed austenite is called retained austenite. The consequence is that the global hardness is limited though the intrinsic hardness of martensite increasing with carbon content. Retained austenite is considered as harmful.

To completely transform austenite, a process with several cycles is recommended. The first stress relieving treatment after quenching, at a temperature of around 190-200°C precipitates very small coherent carbides. As a consequence the lattice distorsion characteristic tetragonal lattice unit decreases in the carbon-depleted martensite. Hence the stress exercised on the nearby austenite decreases; this latter, once released, can be transformed into martensite during new quenching. Every subsequent heating softens the late formed martensite and makes the transformation of a new proportion of remaining austenite possible. The operation repeated two or three times ends in the total transformation of austenite. However, the phenomenon is complex because during each heating process carbon rejected in retained austenite diffuses and takes part in coarsening the small precipitates in martensite. This process leads to a good combination of strength and toughness.

8.5 Solidification structure

Grains, dendrites

A grain is a single crystal, within which the atomic lattice and its orientation are continuous. Adjacent grains of the same phase with different orientations are separated by a surface called a grain boundary. In most metals, the change in orientation across a boundary is abrupt and affects only a few atomic rows, typical boundary widths being of the order of 0.1 to 1 nm. The atoms in the two crystals arrange themselves to prevent the formation of macroscopic cavities.

Solidification can form grains with the state of a perfectly homogeneous and compact crystal as it is common to observe some in mineralogy. The conditions of very slow cooling necessary to obtain such grains are seldom reached in the metallurgy of steels. Then solidification is dendritic. A dendrite forms from a single nucleus. It has a main trunk or primary arm and branches to form secondary and tertiary arms. It results in a rigid structure like a chandelier with multiple arms arranged following a characteristic geometry and gives rise to a grain (Figure 8.5.1 top). It is a single crystal because arms are in continuity with the crystal lattice, so that when all the arms join the whole appears as a single grain. If the sample in Figure 8.5.1 bottom was leveled and polished, only the boundary between the two grains could be observed.

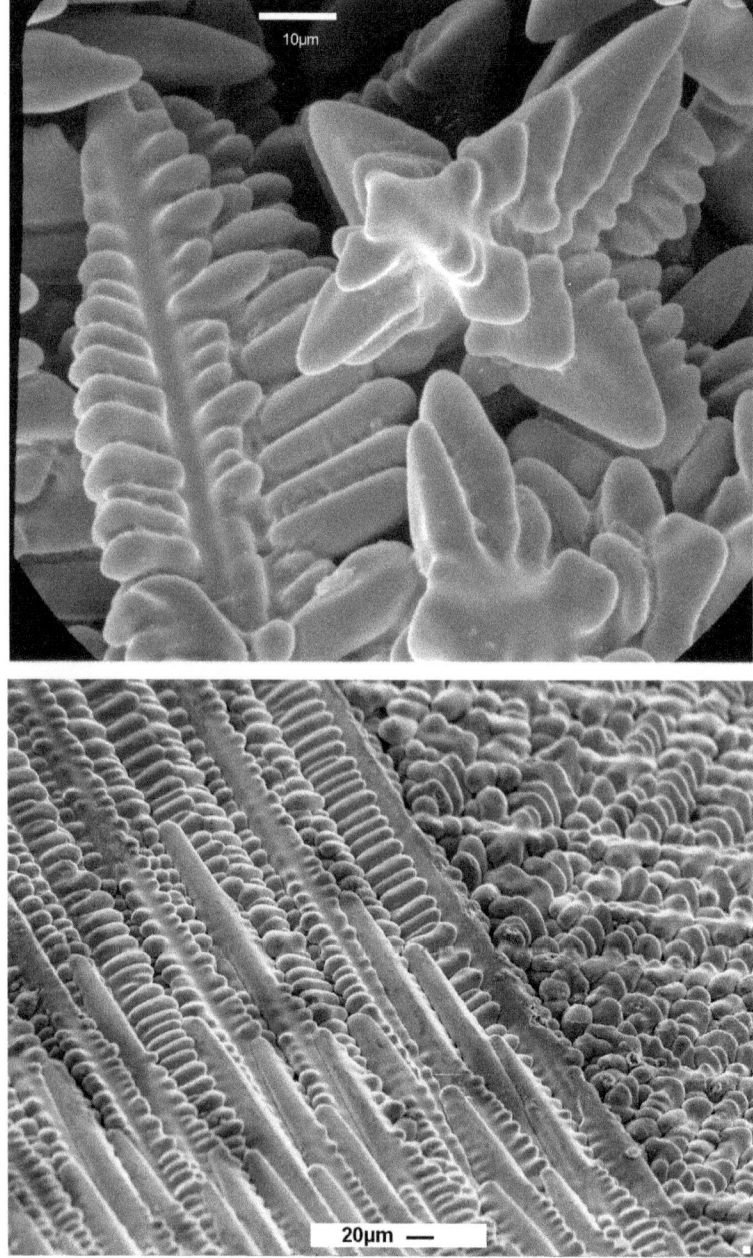

Figure 8.5.1 : Scanning electron micrographs observed in the shrinkage cavity of an as-cast 100Cr6 steel bar. Top : Dendrites with secondary and tertiary arms. Bottom : Two grains, the dendrite arms are parallel inside each grain, since they have formed from the same nucleus. Document Grenoble INP, Fr.

The beautiful organization of the dendritic structure degenerates as soon as it has formed. Indeed, the very active diffusion in the liquid makes that the side branches (known as secondary arms) coarsen, grow and finally separate from the main axis. One can observe this ripening on Figure 8.6.1. Ripening lasts for the time during which the liquid still surrounds dendrites, that is to say the time necessary to cool from the temperature at the start of solidification until that at the end. It is thus the cooling rate in the high temperature range which is determining.

The solidification rate determines the primary dendritic arm spacings and also the ripening of the secondary arms. However the amplitude of the variations is limited. The laws of dendritic growth indicate that spacings are roughly proportional to the cubic root of the time of solidification. In practice, the values vary from several hundreds of micrometers for a slow rate, to some tens of micrometers for a fast rate and some micrometers for a very fast rate in the case of atomizing the liquid metal into droplets for example.

8.6 Dendritic segregation

Non-homogeneous distribution of elements in the matrix

In the metallurgical meaning, segregation is a non uniform distribution in a given homogeneous phase. Taking the example of salt dropped and dissolved in a glass of water, the liquid is considered as a *homogeneous phase*. If it is not stirred, the water is more salted at the bottom of the glass; there is segregation of salt. More scientifically, it is a concentration gradient.

Dendritic solidification involves such a segregation. When dendrites of ice are formed in salted water, the ice is less salted than the remaining liquid and, as solidification progresses, both the ice and the liquid grow richer in salt. As the diffusion of salt in the solid ice is too slow, a gradient of concentration remains between the first formed and the last formed layers, *i.e.* between the center and the circumference of dendrites. Each level of temperature corresponds to a specific partition of every element between the liquid and solid phases.

The consequence of segregation in the formation of a banded microstructure is extensively discussed in the chapters 10 and 11. Segregations are created during the solidification of liquid steel, their formation is illustrated by the diagram showing the dendritic growth during the cooling process under a thermal gradient (Figure 8.6.1). Such images are the result of rather sophisticated computation, though in this case simplified by the fact that solidification starts directly by forming austenite. The images present two stages of the solidification of a bar in a thermal gradient.

Colorings of the image vary according to the local concentration of a given element. The left-hand column shows the austenite growth in the liquid, the central column illustrates the distribution of carbon and finally the right-hand column illustrates the distribution of manganese. Colorings evolve from blue to the yellow when the respective concentrations increase. Thus, it appears that the first solid, at the forefront of the dendrite is depleted in carbon and in manganese compared to the liquid. When solidification progresses the formed solid becomes less depleted in contact with a richer liquid. Rejected carbon and manganese concentrate in the interdendritic space.

In the case of carbon the distribution in the solid becomes homogeneous during the solidification process, its color is uniform. In the case of manganese the enriched layers pile up the ones on the others keeping the memory of the contour of the solidification front as the gradations of colors show it. The explanation is that the considered elements diffuse more or less quickly according to their nature (carbon, a small atom which migrates through the interstitial sites or other metallic elements) and the phase in which they diffuse (liquid, ferrite or austenite). Carbon which diffuses quickly reaches an almost homogeneous distribution in the solid throughout solidification, whereas the manganese which diffuses much more slowly remains segregated (Figure 9.1.5 and Tables 8.1.2, 9.1.6).

The solidification rate has little incidence on the segregations because of compensated effects; in the case of a low rate, diffusion operates for a long time but then dendrite spacings are large and elements must diffuse at a long distance. To summarize, it is practically impossible to avoid a certain amount of segregation at the end of solidification.

UNDERSTANDING STEELS 147

Figure 8.6.1 : Dendrite growth in a thermal gradient for a steel Fe-1C-1Mn as calculated from a numerical model

The cooling rate is relatively slow at about 850°C/h. Dendrites grow in the liquid from a solid nucleus. Two steps of solidification are presented: at the beginning (top) and at about 50% of completion (bottom).

Left column: distribution of the liquid (red) and austenite (white)

Central column: distribution of carbon, lower concentration in blue, read the atomic content on the ruler. To be noticed that at a given level (*i.e.* a determined temperature) the color is uniform, distribution of carbon is homogeneous.

Right column: distribution of manganese, lower concentration in blue, read the atomic content on the ruler. To be noticed that there is a gradient of color corresponding to a manganese gradient in the dendrite.

Courtesy Micress, Aachen, Germany

Macrostructure of ingots

In the case of a massive ingot, cooling is not uniform between the center and the border in contact with the mould. In the outer region the dendritic grains are formed radially, normal to the ingot surface. Inside in the central zone, the solidification is slower, the structure is coarser, more degenerated with randomly oriented equiaxed grains. Many parameters can affect the formation of macro/microstructure of which the most significant is the chemical composition of steel. In short, the microstructure can present oriented grains at the border and coarse randomly oriented grains in the center [70].

8.7 Steels used for cutlery

Optimization of composition, or design

Stainless steels were developed after the second world war when it appeared that an addition of chromium made steel less sensitive to corrosion provided that the proportion exceeds a threshold of about 12%. For the objective of use in knifemaking the optimization of composition to obtain a good resistance becomes a problem because two other qualities should be associated: hardness and good ductility; in other words, the blade must keep a good cutting edge and not break.

To reach hardness the essential prerequisite is to be able to transform austenite into martensite in a certain range of temperature. In this respect chromium is a poor design because it decreases the austenite phase field extent, therefore the possibility of forming martensite. The volume in which austenite exists alone is represented in a perspective view of the ternary phase diagram Fe-Cr-C figure 8.7.1. The addition of chromium markedly changes the extent of the austenite field, which disappears completely at 19%. For higher contents the presence of austenite is still possible but is associated with weakening carbides.

The addition of alloying elements also makes it possible to harden the matrix. Unfortunately the elements which play the best hardening role, like molybdenum and tungsten, reduce even more the field of austenite. The answer consists of adding elements known as austenite stabilizers like nickel known to enlarge the austenite phase field. Modern steels result from subtle optimization.

More recently, nitrogen-containing stainless steels have been developed in which nitrogen is an alloying element equivalent to nickel. Although these favorable actions have been known for some time, the exploitation was limited by the difficulty of introducing large amounts of nitrogen in solid solution. Modern powerful processes of steelmaking under pressure are necessary. These steels tolerate tempering at higher temperatures than those used for carbon steels. But, just like carbon, nitrogen lowers temperature Ms and austenite is transformed with difficulty into martensite.

Brief outline of the ternary phase diagram Fe-Cr-C

The common practice is to represent phase diagrams as a function of temperature and weight or atomic percentage of each element. One of the elements is not taken into account as an independent variable because the sum must be equal to 100%. For example for an iron-carbon steel, the diagram is represented as a function of temperature and of carbon concentration. With three elements the representation requires three dimensions which one can imagine in space. With four elements it jumps into the fourth dimension. Not easy! Then one uses projections on a plane (Figures 8.7.2 et 8.7.3), *i.e.* the variation of only two parameters is considered. However the lever rule usable for a binary diagram is not applicable any more in the same way.

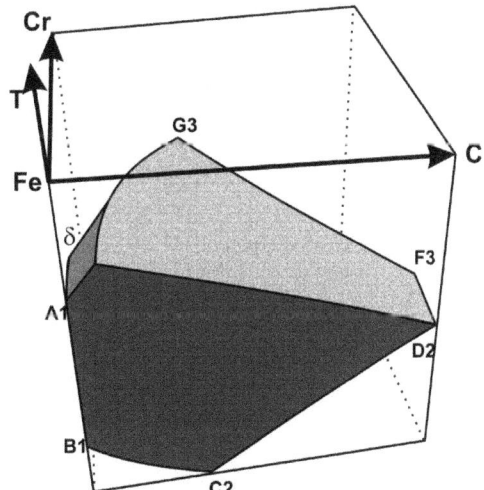

Figure 8.7.1 : Three dimensional representation for the phase field of austenite as a function of iron, chromium and carbon content and temperature. Beyond the limiting surfaces another phase coexists with austenite.
The maximum carbon content is D2 with 2.14%

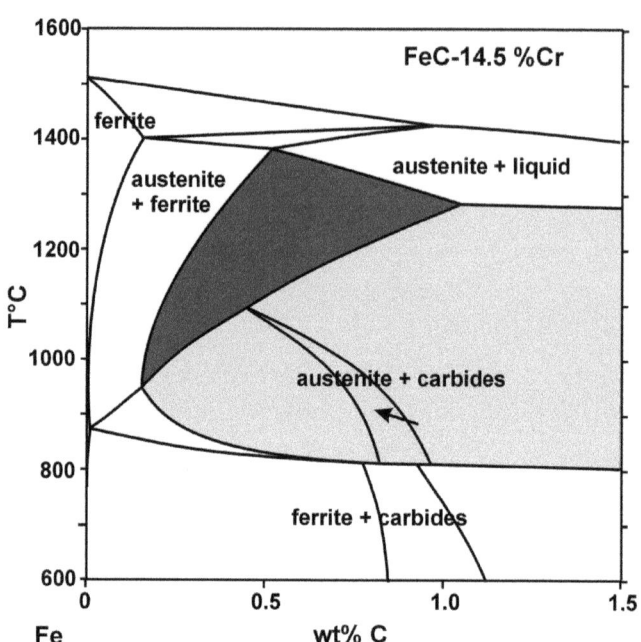

Figure 8.7.2 :
Isopleth of the Fe-Cr-C phase diagram for 14.5% Cr content; the austenitic phase field is significantly reduced, it spreads from 0.1 to 1 % C instead of 0 to 2.14 % C in the case of the Fe-C system.

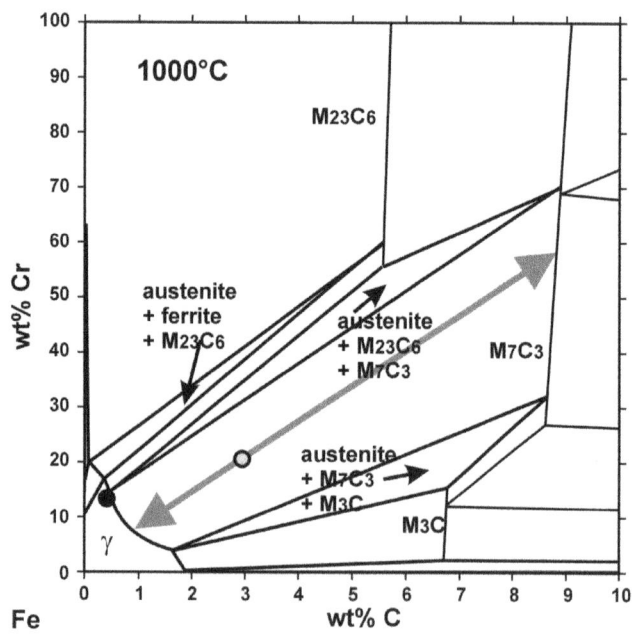

Figure 8.7.3 :
Isothermal section of the Fe-Cr-C phase diagram at 1000°C.
- black dot : 0.6 %C and 14.5%Cr;
- white dot : 3 % C and 20% Cr;
the grey line is the tie-line joining the compositions of the phases in equilibrium for this composition.

To specify, let us consider a section of the diagram with three dimensions, for a chromium content fixed at 14.5%, classical content of the stainless steels (Figure 8.7.2). By comparison with the Fe-C diagram, the field of austenite in dark grey appears very restricted and moves towards high temperatures. If the treatment of austenitization is operated in this zone of temperature, the alloy will be completely transformed into austenite. If the treatment is operated in the light grey zone the alloy will comprise austenite and mixed carbides of chromium and iron. The disadvantage is that a treatment in this window will never re-dissolve coarse solidification carbides, consequently a treatment of austenitization at higher temperature is needed.

The second section was established for a defined temperature of 1000°C (Figure 8.7.3), this section is perpendicular to the previous one. Two usual compositions were reported: the first represented by the black circle is the alloy with 0.6%C and 14.5%Cr. The position of the point in the border of the austenite field shows that the temperature is just sufficient to redissolve solidification carbides. The second white circle indicates an alloy with 3%C and 20%Cr (ZDP189). Its position inside the field where both austenite and M_7C_3 carbide phases coexist, shows that the dissolution of carbides is impossible at this temperature.

8.8 Optimizing microstructure

The microstructure of steel used for knifemaking

A typical example of modern steel is the one labeled X105CrMo17 (1.4125, 440C, S44004) which contains chromium and molybdenum (0.7% C, 1.5%Cr, 0.9% Mo). This range of compositions requires extreme conditions of processing, particularly for heat treatment which results in a high performing steel if they are strictly controlled.

Such alloyed steel forms during solidification of coarse eutectic carbides which can be harmful for the mechanical behavior as well as for the resistance to localized corrosion. The industrial processing must begin by an austenization treatment at a temperature high enough to dissolve chromium and iron mixed carbides. The later processing of hot working of the steel during cycles of forging induce secondary carbide precipitation which are approximately ten times finer than solidification carbides.

The sample examined in Figure 8.8.1 underwent an austenization followed by a cooling during which several types of precipitates were formed :
1- at the beginning of the process, in the highest temperature range, the carbides precipitate at the grain boundaries. As the diffusion is very active they quickly grow and reach a size of several micrometers.
2- several tens of degrees lower the carbides precipitate inside the grains and they are smaller by an order of magnitude.
3- finally below the Ms temperature the matrix is transformed into hard and brittle martensite.
4- a subsequent heat treatment close to 220°C causes the formation of very small precipitates in continuity with the network of the matrix as indicated in § 8.4 (Figure 8.8.1).

The difficult case of steels with high carbon and high chromium content

In the case of steels containing more carbon and alloying elements, the dissolution of eutectic carbides cannot be completed. To avoid a disastrous effect on the mechanical properties the answer is to use the powder metallurgy technology which makes it possible to form alloys in which solidification carbides remain extremely fine.

Nevertheless, the choice of the austenization temperature is still a difficult compromise. The higher the austenizing temperature is, the larger the carbon content of the matrix and, consequently, the lower the Ms temperature. Therefore the limited transformation into martensite leaves a significant amount of retained austenite. Then arises the problems evoked in § 8.4, in particular britttleness. Moreover, as there remain undissolved carbides, these latter are likely to grow easily during the austenization. Conversely, a low temperature of austenization allows the formation of a higher proportion of fine secondary carbide. Quenching is easier but the martensite is softer on account of a lower carbon content.

These considerations are applicable to the steel ZDP 189 with 3% carbon and 20% chromium for which the manufacturer chose the best performances by recommending an austenization at high temperature between 1000 and 1050°C followed by air or oil cooling, still followed

Figure 8.8.1 :
Scanning electron micrograph of an etched sample of a grade for knifemaking (0.7% C, 14.5% Cr, 0.9% Mo) showing complex chromium-iron carbides formed during the cooling process after austenisation.
Courtesy Grenoble INP, Fr.

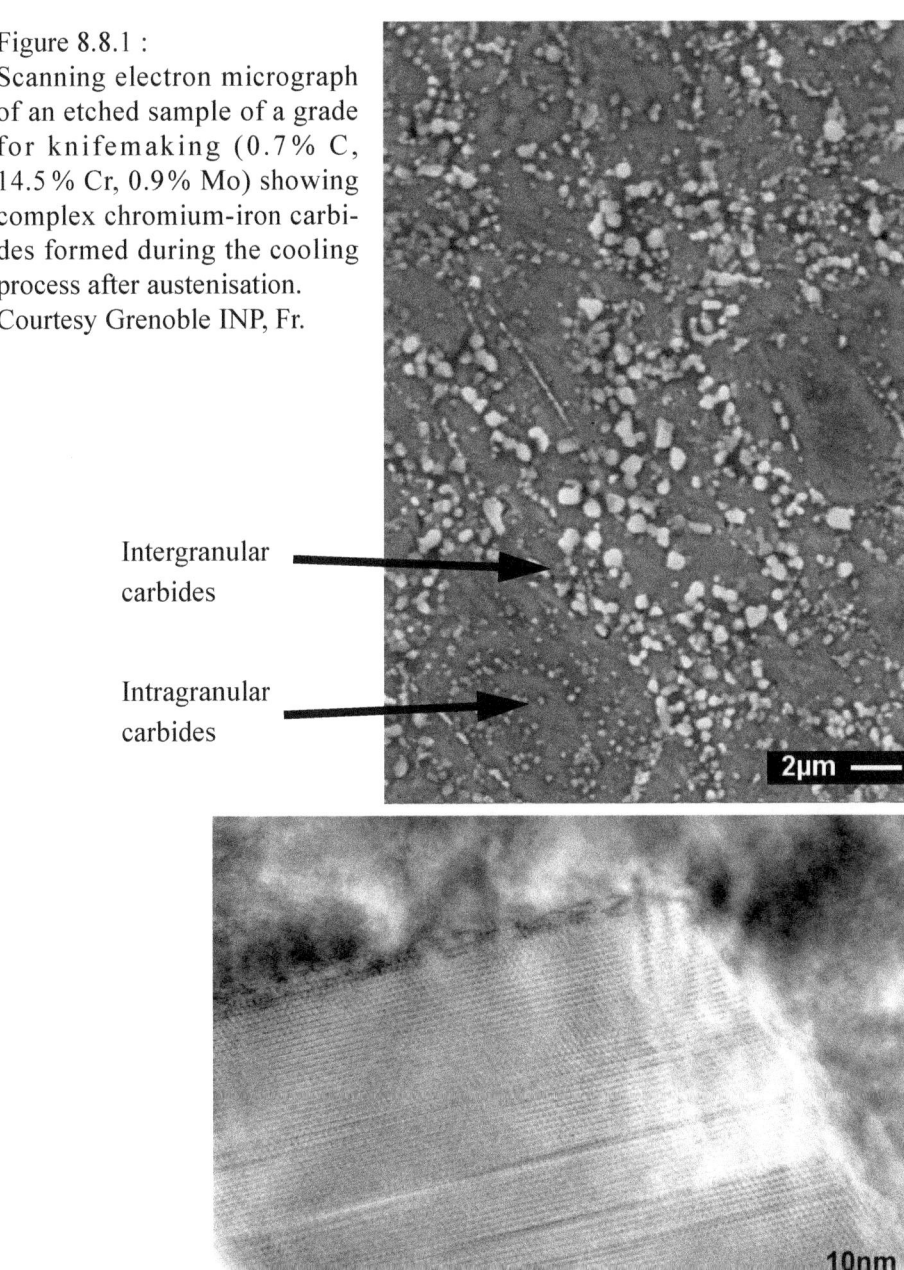

Intergranular carbides

Intragranular carbides

Figure 8.8.1 :
High resolution transmission electron micrograph of a thin foil after stress relieving treatment of the alloy. Visible rows are representative of atomic row.
Courtesy INPG, Fr. and McMaster University, Hamilton, Can.

by a cooling below 0 °C, intended to transform all the matrix into martensite. However treatments involving several cycles of tempering/quenching can also be considered instead of liquid air quenching.

A large choice of grades for cutlery

This is a difficult choice considering the number of commercial grades; here are some families :
- hardenable steels with medium carbon content (0.4-0.8%) in which the additional elements are present at low content to harden the solid solution (Mo), to improve hardenability by quenching, to avoid the graphitization (Mn)
- nitrogen steels which are appreciated thanks to excellent mechanical properties and a good corrosion resistance.
- high carbon steels (0.95-1.2%) with additional elements
- martensitic stainless steels with a high chromium content (13-18%)
- alloyed high carbon steels prepared by the PM technique (§7.5, 8.10).

8.9 Coloration of stainless steels

Some recipes of coloring have been known for a long time, the most famous is gilding mentioned in §6.3 which can be practiced on numerous metals such as iron. Other methods known as burnishing or bluing are specific to the steel which they color in brown or in purple blue. They consist of a chemical attack and formation of a surface layer of colored complex salts. They have a common feature with mercury guilding, it is the use of heavy metals: cadmium and antimony whose salts emit toxic vapors during their heating and which generate polluting waste. The modern methods consist of a simple controlled surface oxidation.

Paradoxically, the stainless steels are made resistant to oxidation by the presence of chromium, but chromium is very oxidizable. In fact it forms by oxidation of a layer of very stable chromium oxide which is protective because it constitutes a barrier for the penetration of oxidant. This providential layer is repaired by self-oxidation in the case of a scratch damage for example.

On an atomic scale the initial layer of chromium oxide is colorless, but if the process of oxidation continues, several layers are superimposed.

For a range of thicknesses about a few tens of nanometers an interference occurs between the received light and reflected light by the substrate, which gives an intense coloration to steel. As the thickness of the layer increases, the color changes passing by different shades: bronze, gold, red, purple, blue and green. The greatest thicknesses give a brownish color without interest. In the absence of chromium in steel, oxidation forms fine layers of silicon or iron oxides which are also coloring but perhaps less protective than the chromium oxide layers.

Two methods can be easily implemented on a craft scale. The first consists in heating steel in the presence of air. The second method is chemical. It consists in immersing steel in an aqueous solution of chromic and sulfuric acids (2.5 mol/L CrO_3 and 5 mol/L H_2SO_4)[2]. In both cases the experimental conditions of temperature and reaction time depend on steel grade. More recently, several electrochemical methods were developed to color stainless steel panels on an industrial scale [79]. They are applicable to the ambient temperature and also use strong oxidizing solutions. After research to finalize the best conditions, these methods make it possible to control the thickness and the surface quality of the layer well, and even, to a certain extent, to confer it a more compact, harder texture. Other electrochemical methods are practiced in a medium of molten salts.

Thermal oxidation gives a soft coloring, chemical oxidation gives a more intense coloring, but it is the electrochemical oxidation which produces the richest selection of colors with golds, reds and blues. The formed layer is stable, resistant, not toxic, but must be protected from the action of an abrasive or a chemical attack by a chlorinated cleaning product.

The surface quality of the substrate is reproduced accurately by the colored layers. A beautiful brilliant color can be obtained only starting from a perfectly polished surface. The austenitic stainless steels are well adapted for this treatment because they give an excellent polished aspect and they contain chromium.

---

2. Imperatively to follow the recommendations of the treaties of chemistry for the mixtures of acids and water which can be dangerous

8.10 Powder metallurgy

Powder metallurgy (*PM powder metallurgy*) experiences a much development because it offers two opportunities:
- to use it for the manufacture of cheap near net shape parts, in particular parts of a car with complicated shapes, built in low alloyed steel with low carbon content.
- to work out grades with a fine microstructure, impossible to achieve by conventional melting and working routes. It is this second aspect which is interesting in knifemaking.

A powder metallurgy process involves three steps: the manufacturing of powders, their shaping and then sintering.

The powders

There are three common types of powders: elemental powder *i.e.* pure metal, prealloyed powder and atomized powder [70, 80-81]. Technically, there are many routes to obtain a metallic powder, however the characteristics of the metal to be prepared allows a limited choice. For instance, crushing is possible only for brittle metals and alloys such as chromium or manganese, excluding iron which is much too ductile.

Iron powder, called carbonyl iron, can be obtained by the decomposition of gaseous $Fe(CO)_5$, in the form of very fine roughly spherical particles, a few millimeters in diameter. The same method is used to produce nickel and cobalt powders. Direct reduction of iron ore is another method which produces coarse, irregular and porous particles justifying the name of sponge iron powder. The average particle size of powder, usually screened to below 212 µm, is about 80 µm. Another more recent process can give a fine powder by forcing filtration of liquid iron

Pre-alloyed powders are prepared by hot diffusion bonding of grains of powders of various elements in the appropriate proportions. The coarse agglomerates of different grains stuck between them look like a metallic muesli.

High alloyed powders such as tool steels, rapid steels or high carbon steels can be obtained by atomizing a stream of liquid metal with high pressure jets of water, steam or inert gas. They are called atomized powders. The liquid metal is broken up into fine droplets which solidify rapidly. The shape of the droplet depends upon the composition of the alloy. In the case of alloy steels, the powders obtained are often coarse and porous. A typical average particle size is about 80 μm. The high solidification speed leads to an extremely fine, generally dendritic, solidification microstructure within each droplet.

Shaping

A binder is generally used to agglomerate the powders to a *green* compact, with sufficient strength to allow subsequent handling. The wax-type binder can be later eliminated by heating. Let us mention the different process of *Injection Molding* in which the mixture binder/powder is a paste sufficiently fluid to be shaped.

One should not confuse this last process with another type of injection molding which takes place during solidification. Molten steel is electromagnetically stirred during the cooling process. This stirring breaks and refines the dendritic structure in the course of formation. When the mixture reaches the consistency of a sorbet, still plastic, it is shaped by molding. This process has the advantage of producing a homogeneous structure, without orientation texture. However the size of the dendritic structure is much coarser than in the case of atomized alloys.

Rapid solidification causes the formation of martensite. The powder must be softened by heat treatment before cold pressing.

Sintering elemental powders or agglomerates

Sintering consists in heating under pressure the powder grains which are welded by diffusion. In the case of low alloy steels the operation takes place at temperatures below 1000 °C for which the powder grains remain completely solid. The holding temperature must be sufficiently long, about one hour, to weld the whole article into a compact structure. The microstructure of the solid thus obtained is fine.

The sintering of the agglomerates leaves a very heterogeneous microstructure. Time is too short and the temperature too low to reach an equilibrium state, for example, between very fine tungsten grains and iron sponge particles ten times coarser. The material retains islands of pure ferrite which are too large to allow significant diffusion of the alloying elements. At this scale the heterogeneity of the material is regarded as a quality because it confers a better ductility.

Sintering atomised powders of alloyed steels

The process is different in the case of atomized powders of high alloyed steels. Sintering is operated at a relatively high temperature, higher than 1000 °C and it is accompanied by the formation of a small proportion of eutectic liquid between the dendrite arms. These powders are coarse but within each micro-ingot, corresponding to a solidified droplet, the dendritic structure is very fine with a spacing of about a micrometer and is practically deprived of secondary arms. The eutectic interdendritic zones are also very fine, submicron resulting in tiny eutectic carbides.

Segregation does not disappear, but it is miniaturized. In fact, the microstructure is so fine that the equilibrium state is readily established in alloy during a further holding in temperature, contrary to the alloys formed with agglomerates. It is important to control very strictly the sintering conditions: control of duration to limit a possible slight coarsening of carbides, and control of temperature to adjust the proportion of liquid which must remain lower than 20%. The range of appropriate temperatures, said to be the sintering window, is often rather narrow and is situated above 1000 °C in the case of steels. It is optimized for each grade. The powders used to produce damask steels Damasteel® are atomized powders. Both operations of shaping and sintering are merged when the powder is hot pressed under high pressure. A powerful method HIP (*Hot Isostatic Pressing*) consists in heating the powders under an isostatic pressure of about 100 MPa.

Two methods comprise a sintering under hot isostatic pressure: the *Damasteel®* process, [73], § 7.5 and Figure 7.5.3 to produce directly laminated steel and the process called *Crucible Particle Metallurgy processing* (CPM)® to produce high alloy steels. The latter should not be confused with the preparation of wootz also called crucible steel for which there is fusion of the steel and solidification inside the crucible.

3D printers are used successfully to print steel objects. The technique consists in monitoring a powerful beam in order to fuse a small drop of metal; either a gas-metal arc welder to lay down thin layers of steel or a laser beam which fuses metal powder. Laser sintering of gas atomized metal powder enables deposition of specially designed alloys. 3 D printing will probably open new types of pattern inlays for blade decoration.

9 *Pattern-welding*

The historical part of this work showed how the oldest welding method was used to manufacture iron parts starting from smelted iron. The main problems are the quality of the weld. Modern means of observation provide an objective answer.

9.1 Welding different layers

Welding tracks

Ore smelting produces an iron bloom which must be hot worked to eliminate the most of the residual oxides (§ 1.2). The smith works by hammering and, when the metal is highly thinned, he welds piles of strips similar to the preparation of puff paste. In French, the metal is said to be *corroyé* (curryed) which means beaten in bundles or sheaves. In the metallurgical context, we mean hot working of several sheets or rods between 1000 and 1100°C.

According to the conditions of heating and air blowing, the surface of the metal can be carburized (Figure 2.2.4) or decarburized because of an oxidizing atmosphere or the presence of flux (Figure 9.1.1). These surfaces put together in contact keep a visible print of the welding, at least when using a microscope. Oxides entrapped between the metallic layers form alignments of inclusions. These tracks show the way the metal was forged and welded. The welds of samples from the first millennium are more or less coarse and unrefined (Figure 5.2.2) and there are zones in which rupture is easier since cleavage fracture usually occurs at locally brittle points such as inclusions.

Each folding doubles the number of layers and divides their thickness by as much, so that ten operations starting from layers of 1cm would lead to 1024 layers of 10 micrometers! In fact, the oxide coatings split and spheroidize during heatings while the matrix is homogenized, more or less completely, by carbon diffusion (Figure 9.1.1 and see also [82]). A few elongated precipitates correspond to the last forge welding.

Figure 9.1.1 :
Distribution of inclusions in a steel :
Top : Optical micrograph of a sample after eleven operations of forging then folding using borax. Clear lines correspond to decarburised zones near the welds.
Bottom : Electron micrograph in which oxides appear as small droplets in clear contrast. Courtesy Grenoble INP, Fr.

The fact that oxides do not constitute a continuous barrier any more is important for the future mechanical resistance of the blade.

The goal of the blacksmiths in the first millennium BC was probably not to produce banded structures but to homogenize, to refine the steel with the aim of working out a more resistant material (example § 6.7).

Borax

The best coating used to protect forged parts is borax because of its exceptional wetting ability and its reactivity with respect to oxide layers. It spreads out perfectly over the metallic surface and forms a very fine layer resulting in a smaller proportion of oxides entrapped along the welded joint.

The compound called borax is a mixed oxide called sodium tetraborate, of chemical formula $Na_2B_4O_7\text{-}10H_2O$. This salt is water soluble, this is why it can be found essentially in arid areas formed by evaporation of salt lakes. The name is recorded since the 9th century but also was known by the name of *Baurach, Baurak, Boras, Tincal* and finally *Chrysocolla* (Agricola). Exposed to dry air, it is transformed into a chalky white material. The variety dehydrated with half of the water is called *Tincalconite*. The first borax specimens came from deposits in dried up lakes in Tibet. They were conveyed in great quantity by caravans. The modern supply comes from the American continent: Argentina, Mexico, the USA (California). In antiquity, a source liable to be used, in addition to Tibet, was Turkey.

Borax has many remarkable features, it is recognized as a detergent and having disinfecting properties. The molten borax forming a flux able to dissolve metallic oxides is extremely useful to realize welds and to work iron. It was most probably used for a very long time in the Middle-East. Because of its great solubility in water it is not possible to find it any more on the ancient sites of forging mills or ovens. The track of its use can be proved by analysis of inclusions. However, boron is a light element, difficult to analyze acurately, which requires a specific procedure with the electron microprobe.

The contemporary blacksmiths often use borax by powdering the part during forging (Figure 9.1.2). The hydrated compounds lose their water under the effect of the high temperature producing a kind of bubbling, that is why the anhydrous quality can be preferred. The residual borax also may damage the furnace walls by chemical attack during heat treatments.

Figure 9.1.2 : Vulcain's secret, acrylic painting.
Courtesy Sonia Durand, Fr.

Phosphorus in ancient steels

In the case of Merovingian swords, the welding of alternate layers from different grades is deliberate and different compositions are chosen in order to form a decoration. The layers in general consist of low carbon steel, rich-carbon steel and low carbon steel containing some phosphorus.

The problem is then to avoid homogenizing the welded unit. Landrin [52], gives advice to knifemakers : *It is especially necessary to be careful not to heat too much : the beauty and the goodness of the blades consist mainly in this: it is necessary that each matter that one employs be preserved without being denatured. It is necessary that steel keeps its quality and the iron its: too strong heatings would confuse them together.*

Phosphorus, an element naturally present in many ores (for example the iron ore of Lorraine called *minette*) is frequently detected in one of the layers up to a proportion of about 3.5%. Phosphorus facilitates welding by lowering the point of onset of melting by formation of a eutectic liquid as low as 966 °C instead of 1127 °C in the case of Fe-C alloys. However, phosphorus is regarded as a weakening element. This last effect can be avoided by alternating phosphorous layers with layers of pure iron. Phosphorus also inhibits the diffusion of carbon between the layers and avoids a too fast homogenization during multiple foldings.

The sample shown in Figure 9.1.3 corresponds to the Merovingian sword of Figure 3.1.4 which is constituted by layers of two alloys among which one (bottom of the micrograph) contains approximately 0.3-0.4% carbon taking into account the proportion of pearlite, and phosphorus with a content less than 0.1%. The other layer (top of the micrograph) contains less carbon but has a rather high phosphorus content which was estimated at about 0.6-0.7 wt%. In the P-rich upper part, the grains are surrounded by a kind of river with a smooth contour and elongated carbides in the middle (see magnification). This microstructure suggests that the steel was melted in this groove and the liquid formed a very fine eutectic component during the cooling process. For contents ranging between 0.1% and 0.3% of phosphorus,

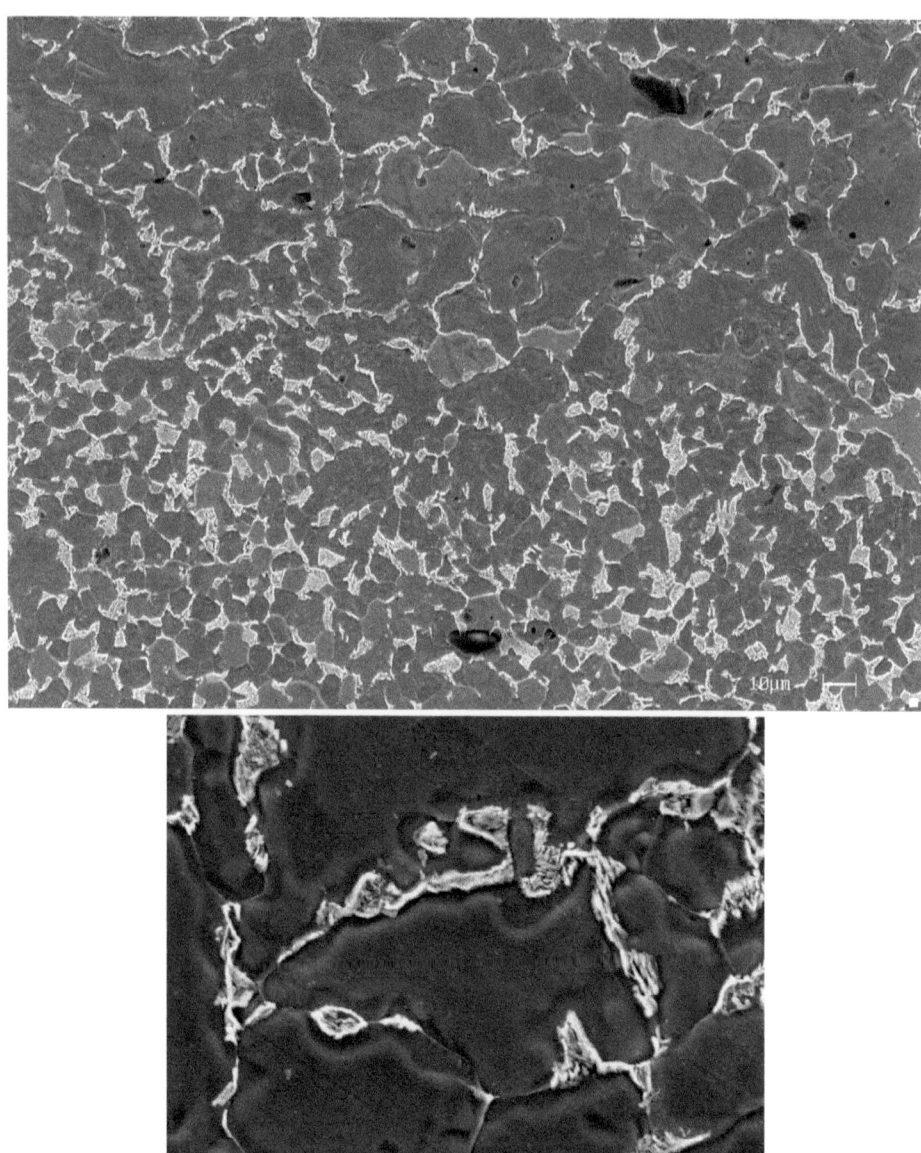

Figure 9.1.3 :
Top: Electron micrograph showing that the transition between the two layers is progressive and large (about 50μm).
Bottom: Magnification of a grain in the P-rich layer illustrating the structure of the remelted groove. Courtesy Grenoble INP, Fr.

PATTERN-WELDING 165

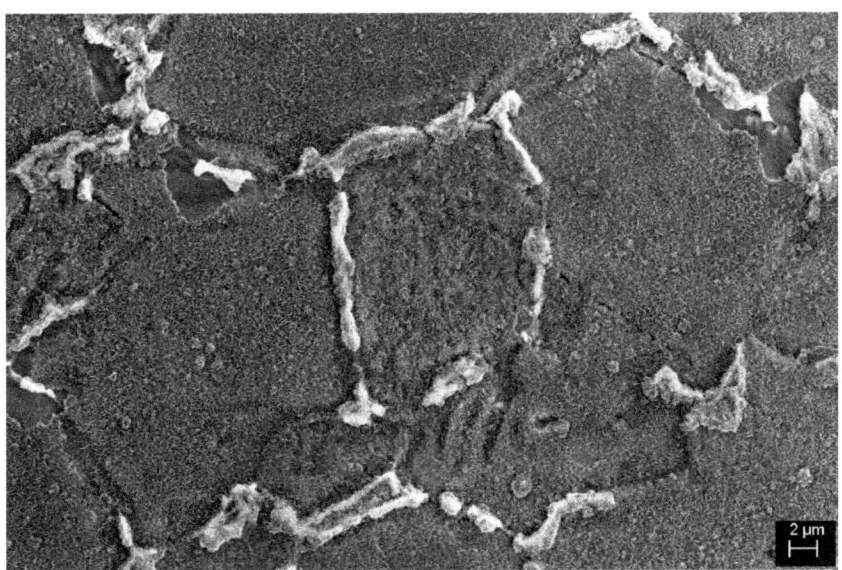

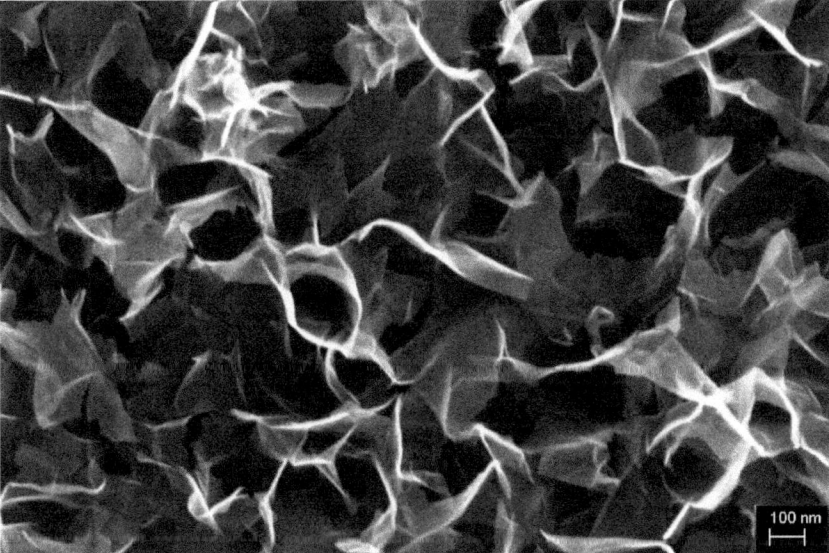

Figure 9.1.4 : Electron micrograph of the sample etched with nital
Top: a similar area as in Fig. 9.1.3
Bottom: high magnification illustrating the particular structure.
Courtesy Grenoble INP, Fr.

fusion can occur locally at the grain boundaries where carbon and also phosphorus segregated.

A similar area of the same sample was observed after a deep etching with nital (Figure 9.1.4 Top and also [83]). The micrograph appears fuzzy and blurred. This observation was often reported for steels containing phosphorus. The fuzzy aspect is due to the relief created by chemical attack, which is why this structure called ghost structure. This is typical of the ancient irons containing phosphorus. An examination with very high magnification (Figure 9.1.4 Bottom) reveals a microstructure which evokes typically an intergranular fracture topography. However the size of the grains is very small here, of a hundred nanometers, thus rather corresponds to the size of sub-grains.

Phosphorus atoms migrate to the grain or sub-grain boundaries on account of a higher content of defects such as vacancies or dislocations. This segregation enables formation of a particular bonding between phosphorus and the iron substrate with the Fe_3P structure, a kind of layer limited to a few atomic rows which acts as a barrier against chemical corrosion.

The blacksmiths of the Middle Ages adapted skilfully their process to the disadvantage of phosphorus which is brittleness. They made this defect less catastrophic by alternating phosphorous iron with ductile layers of almost pure iron. Moreover, they exploited some advantages: a slight hardening, an increased resistance to corrosion and easier welding.

The effect of nitrogen

Nitrogen is, like carbon, a powerful strengthening element. The question was asked to determine if its presence at a low content in old blades was the result of an enrichment carried out by carburization and nitriding of iron [84]. Was the former use of die coats incorporating magic ingredients justified? An attempt to nitriding was practiced in a completely empirical way, by heating iron with coal and nitrogenous organic waste such as droppings and manure, known to be suppliers of carbon and nitrogen. But, in these conditions the incorporation of nitrogen into iron is difficult and limited to very low contents. Consequently, the contribution to hardening proved to be tiny. The

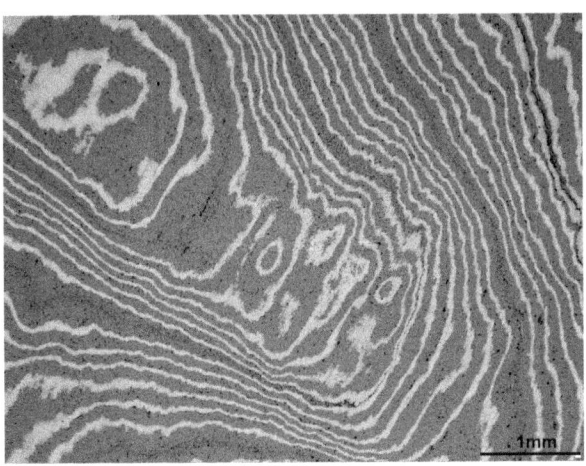

Figure 9.1.5 :
Electron micrograph of a laminated steel obtained by forge welding three grades XC90, XC45 and a steel with 4% nickel.
Experimental specimen. Courtesy Henri Viallon, Thiers, Fr.

knowledge which was transmitted undoubtedly originates more in legend than in know-how.

This recalls the legend of Wayland (German, Wieland), smith, artificer and king of the elves in ancient European folklore dating back the 12th century, who was dissatisfied with the first forging of his sword Mimung and broke it into thin fragments which he mixed with flour and fed to ducks and geese. Regretting his act, he recovered the metal in the birds' excrements and found the oxides to have been cleaned away. Then, he forged the metal together with the dung, repeating the operation several times, and obtained a sword of incomparable quality. Scientific experiments in 1930 demonstrated that heat treatment in nitrogen-rich bird droppings can effectively slightly increase the nitrogen content of iron [8].

Ancient blacksmiths adapted their forge-welding process to the available steels. Modern blacksmiths have a wide choice of grades free of phosphorus or other harmful elements. The welding process does not usually involve the formation of liquid between the layers. Examples are presented in §7.3. However elements diffuse more or less rapidly through the interfaces tending to homogenize the layers. Figure 9.1.5

Table 9.1.6 : Features of forge welding for different grades

Composition of layers	Homogenization	Microstructure
Different only by carbon contents	Rapid Risk of graphitization when forged for too long	Entrapped oxides Spheroidization of oxides, Embrittlement if coarse plates of oxides Figures 5.2.3, 9.1.1
Different with alloying elements (Cr, Ni)	Slow, significantly decreased Hot working more difficult with Cr, higher temperature required Risk of graphitization with Ni	Distinct layers Clear contrast with Ni Risk of rupture along the weld with Cr
Element liable to lower the melting point (P, S)	Welding is easier P inhibits homogenization	Risk of embrittlement with P Risk of hot shortness Figures 3.1.4, 9.1.3, 9.1.4
Liquid layer	In the weld	Progressive transition
Use of borax	Weld cleaner	

presents a micrograph of a forge-welded stack made with three grades. Only two differents layers can be observed because diffusion of carbon has homogenized the two carbon steels. Some features of the weld in relation to composition are summarized in Table 9.1.6.

10 Moire pattern in wootz type, high carbon steels

The wootz is often called the damask of crystallization. It is justified because such a structure is obtained starting from as-solidified crucible steel, nevertheless it is inaccurate insofar as the microstructure which makes the decoration is formed during complex structural transformations occurring after solidification.

10.1 Crucible steels (wootz, pulad)

The ingots of wootz or cake

The so-called crucible steels were prepared by fusion (see §1.2), in the shape of small ingots or cakes. When solidification is achieved in the crucible the structure comprises rather coarse and disordered grains inside which dendrites underwent a certain globularization, a ripening, but not a degeneration which would have completely split them. An example of macrostructure is presented in the book of Figiel [16] and a rather similar modern ingot is shown in Figure 7.7.3 and in reference [74]. The solidification grains are areas of several millimeters wide visible to the naked eye, they are defined by their specific orientation. In each grain the primary dendrite arms are parallel and secondary arms perpendicular to the primary arms.

The cakes are forged in the as-solidified state. The grains are deformed during forging, their contour accompanies the deformation of the matrix. When the deformation is sufficiently intense, the grains split and are rearranged at high temperature, and contours are redistributed, this is the recrystallization. Blacksmiths are able to detect directions in which the deformation is more natural but the macrostructure has little incidence on the final result.

Another essential feature observed in real Damascus steel blades is the presence of a decarburized line of pure ferrite along the dorsal ridge of the blade. Verhoeven [85] proposes an explanation which illustrates the swordsmiths' technological skill. Wootz steel often contains sufficient amounts of phosphorus and sulfur to lower the incipient melting temperature to around 966 °C. The resulting hot shortness causes delamination during forging. In order to overcome this problem, a prior decarburizing treatment was performed, consisting in a holding at 1200 °C for about 5 hours, followed by water quenching. A thin decarburized layer was formed at the surface, with a much higher melting point than the carburized zone. The cake can withstand the first hot hammering through its malleable ferritic envelope without cracking. The dorsal ridge is the only remaining evidence of this process and is due to a single initial folding operation. Forging is pursued by cycles of hammering and reheating.

10.2 Formation of the moire pattern

To understand the formation of this microstructure let us distinguish different steps corresponding to ranges of temperature defined by the Fe-C phase diagram. The transformations occurring, either on heating or on cooling, develop specific microstructural features which enable recognition of their origin.

The first step, I in the phase diagram in Figure 10.2.1, corresponds to solidification which induces segregation of carbon and of dissolved elements (impurities frequently coming from the ore) and results in the formation of *coarse eutectic carbides.*

The second step, II defines a range of temperatures between which the equilibrium state is a homogeneous solid solution of austenite. It means that holding the alloy in this range of temperature induces more or less rapidly the dissolution of all the carbides.

The third step, III corresponds to an equilibrium state between austenite and carbides. During the cooling process austenite undergoes a precipitation of carbides labeled *secondary carbides.*

The fourth step, IV is the range of temperature in which austenite undergoes solid state transformations depending upon the cooling rate: pearlitic, bainitic (upper or lower bainite) or martensitic.

The transformations involved in steps I, II and III are controlled by the diffusion rate of the elements, mainly carbon. The higher is the tempe-

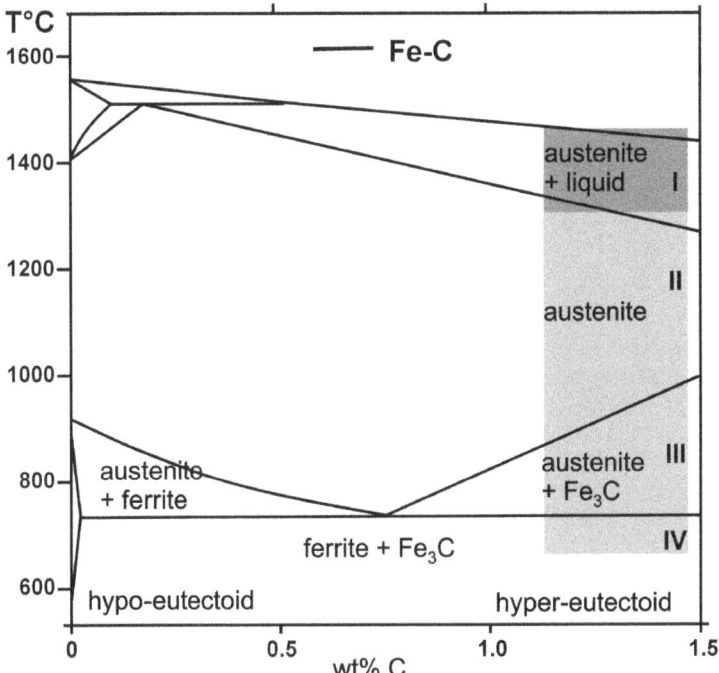

Figure 10.2.1 :
Calculated phase diagram Fe-Fe$_3$C. Courtesy INP Grenoble, Fr.

rature, the easier are the transformations, in particular spheroidization and the coarsening of carbides.

Components which form the pattern visible to the naked eye

Eutectic and secondary carbides are too small to be individually distinguished, however they are discernable when they are grouped in alignments. In the case of high carbon steels such as wootz steel, the moire effect, the decoration with roses or ladder visible at surface are made apparent by secondary carbide alignments. In the case of low carbon steels studied in chapter 14 these alignments consist of a string of pearlitic cells also visible to the naked eye. Carbides formed in step IV are finer than secondary carbides, they are visible only with a good microscope (Figure 8.4.1). As they are regularly distributed they contribute in no way to form a pattern, a texture. Each constituent lies in a range of size for which an approximate value is reported in Table 10.2.2.

The contribution of each one to the visible pattern is clearly pointed out for a scale extending from 0.01 to 10000 µm.

Table 10.2.2 : Approximate sizes in micrometers of constituents which can be observed. In the frame are the only ones responsible for watered or moire patterns visible to the naked eye

Solidification grains	10 000 µm	Dendritic single-crystal with main trunk, secondary and tertiary arms
Recrystallization grains	1 000 µm	After hot working and recrystallization
Dendritic arms	100 µm	Spacings between interdendritic corridors
Eutectic carbides	10-20 µm	Spacings between eutectic plates
Secondary carbides	2-5 µm	Formed in the range 725 - 1000 °C in austenite
Pearlite cells	10-30 µm	Formed below 725 °C
Very fine pearlite cells	2-3 µm	Formed below 725°C (rapid cooling)
Carbides resulting from tempering	1 µm	Formed within martensite 500-600 °C
Carbides, clusters, nanometric precipitates	0,010 µm	Tempering of martensite (200 °C), bainite or slow cooling rate at low temperature

The secret of oriental damasked steels lies in the art of creating these famous alignments of carbides. Even when the western smiths could obtain imported wootz at their disposal they did not succeed in reproducing the damask structure. The rare ones who forged this carbon-rich steel understood that the cake had to not be reheated above bright-red hot during forging. Indeed, the beginning of melting occurred in the segregated zones starting from 1147 °C at the eutectic temperature and even at a lower temperature in the presence of phosphorus or of sulfur. The metal splits and then blows under hammering.

During recent decades the interest was revived because two research teams established different experimental procedures able to reproduce a microstructure apparently similar to those of genuine swords. The mechanisms of transformation, clearly highlighted, are different giving rise to controversy. In fact, they are two academic models which clarifies various aspects of a complex transformation. To understand the behavior of wootz it is necessary to throw oneself bravely into the complexity of high-carbon steels!

Wadsworth-Sherby procedure

The first experimental technique was established by Wadsworth and Sherby from Standford University (USA) in the 1980s, leading to structures apparently similar to those in genuine Damascus blades [58, 86-89].

The essential step is one holding the temperature at around 1093 °C for 48 h. This very long heat treatment has the effect of completely dissolving all the cementite and obtaining a fully homogeneous austenite structure without leaving tracks of dendritic segregation. The alloy adopts, whatever its origin of elaboration, crucible steel or industrial steel, a structure of coarse-grained austenite. The temperature was selected in a region usually avoided in the industrial processes because it also causes grain growth of austenite. The steel is then cooled very slowly and the pro-eutectoid cementite forms more or less acicular coarse precipitates which nucleate at grain boundaries and grow along preferential crystallographic directions.

Finally, in the last step, the billet is worked at dull red heat, just above the eutectoid temperature *A1*. The heavy deformation involved in manual hammering is simulated by a series of large hot rolling reductions. This procedure broke down the coarse intergranular carbide particles into more or less angular fragments, a few microns in size, strung out in rows parallel to the rolling direction. The crushing of carbides is accompanied at high temperature by their partial globularization. It is these coarse carbide alignments that produce the pattern observed in Damascus steel blades. The matrix undergoes the pearlitic transformation, it is transformed into ferrite with very fine carbides, much finer than aligned carbides.

This treatment can be compared to that of modern alloyed tool steels in which the eutectic complex carbides adopt complicated, harmful shapes. A thermomechanical treatment with a strong reduction ratio up to 80% is necessary to transform them into fine globular carbides [70].

Verhoeven-Pendray procedure

After a detailed examination of the published literature, Verhoeven, Pendray et al. [59], an Iowa University team, carefully studied a series

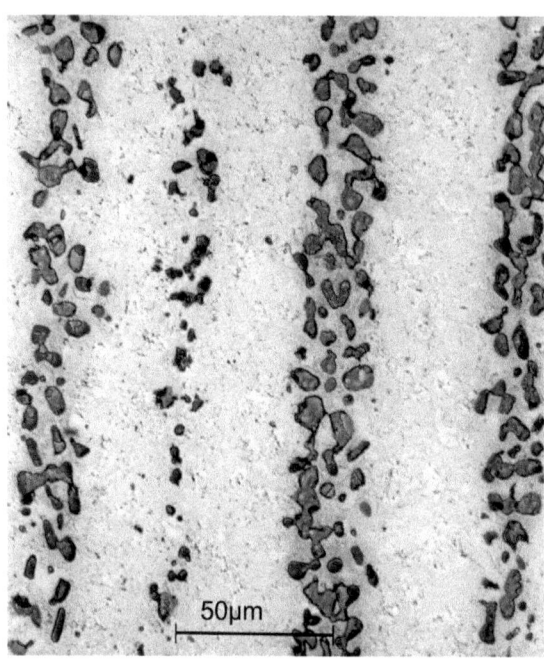

Figure 10.2.3 :
Microstructure of the blade forged by Al Pendray presented in Figure 7.7.1. The carbides of cementite in dark contrast have a mean size of about 6 micrometers.
Courtesy University of Iowa, USA.

of ancient Damascus steel blades and pointed out a number of microstructural features that can be considered as criteria for distinguishing genuine Damascus steels [60, 90]. They noticed a correlation between the spacing of the carbide rows and that of the primary dendrite network in the initial alloy. Dendritic solidification results in precipitation of eutectic carbides but also in residual segregation of impurities. Small, but significant differences in composition were effectively detected between the centers and edges of the carbide rows. The principal impurities concerned were found to be silicon, phosphorus and vanadium in relatively low concentration.

Experiments have been performed on synthetic materials of similar composition. Al Pendray forged a blade shown in figure 7.7.1, the microstructure of which shown in Figure 10.2.3 satisfies all the criteria of comparison established for genuine blades. Again the procedure distinguishes two main steps. The first one is a heat treatment in a range of temperature sufficient to completely redissolve the coarse interdendritic eutectic carbides. In practice, it consists in a hold for a short time at a temperature above the limit of solubility around 1050°C and not exceeding 1200°C. The heat treatment is not long enough to homogenize

Figure 10.2.5 :
Electron micrograph of a vanadium cast iron illustrating the growth of cementite in continuity with vanadium carbides in black contrast, after [69]. Cementite is the clear edge. These carbides are relatively large with 4-6 µm of diameter.
Courtesy Grenoble INP, Fr.

the austenite, the alloying elements other than carbon remain irregularly distributed with a higher content in the previous interdendritic areas.

During the second step the sample is subjected to many cycles of alternating heating and cooling in order to simulate what occurs during manual forging. Precipitation tends to be accentuated selectively in segregated zones during cycles. Indeed, the short heating, just necessary for austenizing, redistributes carbon only at close distances, while during slower cooling carbon is drained towards the precipitates already formed in the corridors of segregation. It was demonstrated that the rows of carbides can be generated by heat treatment cycles, without the need for plastic deformation, either by hammering or rolling.

The key role of vanadium

Verhoeven [59] pointed out that vanadium plays a key role for this controlled precipitation. It can be explained considering phase equilibria. During solidification vanadium partitions between the liquid and austenite resulting in an irregular distribution of the vanadium content in a similar way to manganese in the modeling presented in Figure 8.6.1, the higher concentration being at the border of the dendrites. Thus VC carbides precipitate firstly in the richest zones and before cementite according to the phase diagram in Figure 10.2.4.

These carbides precipitate easily in great number. However they remain very small because the low vanadium global content limits their growth. Their essential role is to be used as a nucleus for cementite at lower temperature. Hammering does not inhibit precipitation

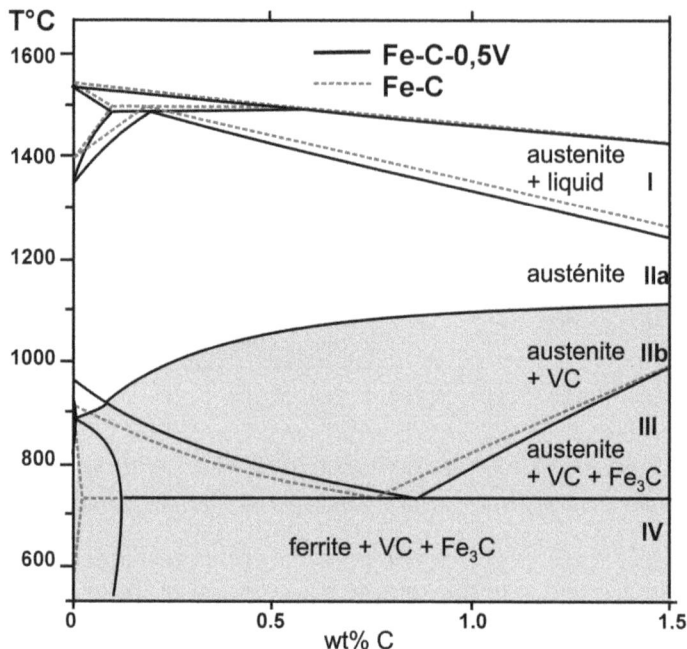

Figure 10.2.4 :
Isopleth of the Fe-C-V phase diagram for a content of 0.5 wt% V showing the phases liable to form during each step of the cooling process.
Step I corresponds to the transformation of liquid into austenite. The solidification of the last liquid results in the formation of eutectic carbides.
The second step consists in a hold in the IIa field for a short time. The limited temperature of this phase field making it possible to dissolve carbides without homogenizing completely the alloy elements other than carbon.
Step IIb corresponds to the precipitation of VC carbides.
Finally step III is the precipitation of cementite secondary carbides.
This value is selected relatively high compared to an initial concentration of about 0.1%V, this is because it takes also into account a significant enrichment in the dendritic borders where precipitation occurs.
Courtesy Grenoble INP, Fr.

and can on the contrary facilitate it by creating defects which constitute additional sites of nucleation.

The role of the vanadium is indeed exceptional!
Srinivasan and Ranganathan [11] call vanadium *the magic dust*. Vanadium is known, together with three other elements, titanium, tantalum

and niobium, for its tendency to form with carbon, small, dispersed carbide precipitates. This aptitude is related to the phase equilibria of the ternary system Fe-V-C: in the high temperature range the solubility of vanadium is lower in austenite than in the liquid metal (partitioning coefficient around 0.5) resulting in a significant segregation gradient from the center to the border of the dendrites. This explains why an initial content of about or even less than 0.1% is sufficient. At temperatures in the forging range, the solubility of vanadium decreases with temperature inducing the precipitation of vanadium carbides [91]. Another aspect is crystallographic, the nucleation of VC is very easy on the network of austenite and once precipitated VC is used in turn as a nucleus for cementite. On account of the presence of similar atomic planes between both structures, cementite can grow in continuity with the crystallographic network of vanadium carbides which is illustrated in Figure 10.2.5. Examination of phase diagrams shows that none of the other elements such as titanium, tantalum, niobium, chromium, molybdenum which were suggested by many authors should be adapted as nucleation agents. It was by chance that the old methods allowed the incorporation of vanadium because it is the element which is best designed in this case. Verhoeven [59] suggests: *the possibility that the low levels of vanadium found in the genuine wootz blades may have resulted from ore deposits in India where the wootz steels were produced.*

But, is this role essential?
Fortunately no. In the absence of vanadium the residual segregation of carbon should be enough to generate precipitates in the enriched zones provided that the distribution of carbon is not homogenized *i.e.* the alloy is not heated too much and for too long. After each new heating process, carbides dissolve and there remains locally an excess of carbon which is not leveled immediately by diffusion. Precipitates of undissolved carbides can be also used as nuclei. Precipitation is however likely to be less easy, less regularly distributed and requiring a higher number of heating/forging operations.

In Table 10.2.6 the main features of each procedure are reported. These models prove that there are at least two possible procedures to obtain carbide alignments. Wadsworth's model seems less credible in the application because of the difficulty of total homogenization, however it highlights the possibility of crushing the carbides.

Verhoeven's model shows the essential aspect, *i.e.* the role of solidification segregations. The initial conditions are strictly defined in both cases, also it is difficult to imagine that thousands of blacksmiths who forged wootz successfully for centuries, in extremely distant countries, all adopted the same procedure without deviation.

Table 10.2.6 : Summary of the procedures

	Wadsworth-Sherby	Verhoeven-Pendray
Step I	complete homogenization, intergranular precipitation during slow cooling	dissolution of coarse interdendritic carbides, non redistribution of alloying elements
Step II	crushing, rearrangment and globularization of intergranular carbides during hot working	heating and cooling cycles in order to precipitate new cementite particles in a controlled manner

What occurs when the conditions are slightly different? The question is asked to determine how the morphology of precipitation can be affected by variations in the procedure of forging. Are the criteria defined by Verhoeven on the distribution, morphology and size of cementite particles always checked? (Figures 7.7.1 et 10.2.3)

The most probable variation to Verhoeven's conditions is that the steel is not heated at a sufficiently high temperature enabling the complete dissolution of eutectic carbides. This assumption is justified insofar as the presence of steadite was detected in certain crucible steels [92]. For such steels the initial heating must be moderate because this ternary eutectic austenite/cementite/Fe_3P has a melting point as low as 966 °C resulting in the formation of a small proportion of liquid (hot shortness). As a consequence blacksmiths, who controlled the temperature only by estimating the shade of red, cautiously use this to avoid excessive temperature likely to break the ingot during forging.

The residual eutectic carbides transform during hot working. While the ductile matrix is deformed, carbides which are not ductile do not deform but break. They are partially split up, dislocated and sheared. The cycles of heating and cooling tend to crumble and spheroidize them by successive dissolutions and precipitations. The driving force of these transformations is the minimization of surface energy. Junctions between the thinnest precipitates and the most protuberant less

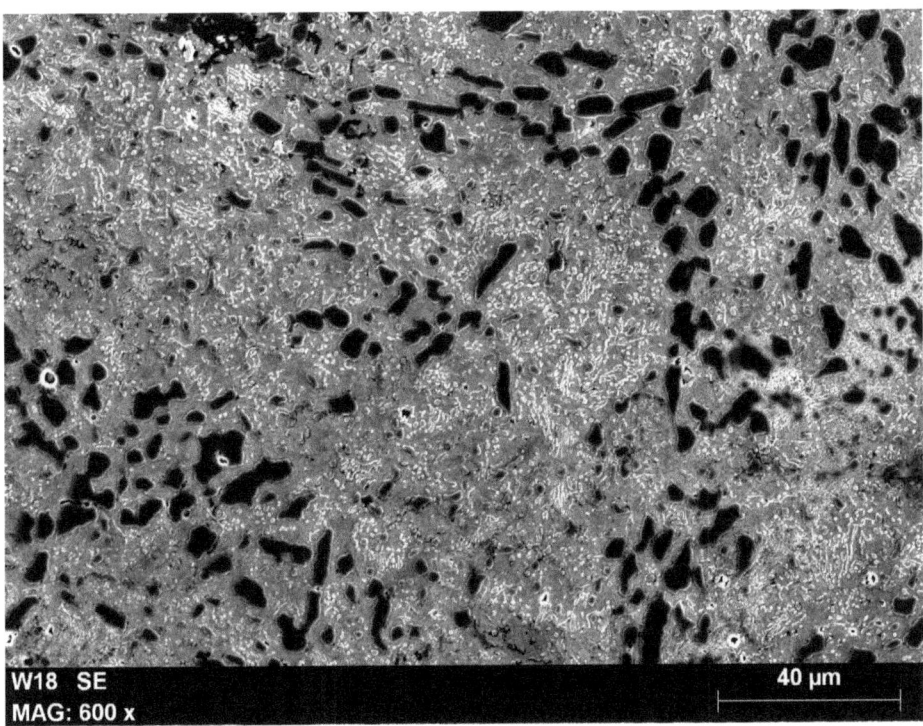

Figure 10.2.7 : Coarse carbides in dark contrast (type 1) are distributed along irrregular rows forming the moire pattern that can be seen with the naked eye. Their size ranges between 4-8 µm which is in agreement with Verhoeven's findings. Typically these carbides are transformed eutectic carbides.
SEM electron micrographs, courtesy CMTC, INP Grenoble, Fr.

stable zones tend to dissolve and at the end of many cycles the initial eutectic morphology disappears to be replaced by more or less regular carbide alignments.

Since the publication of this book in French, the author had the opportunity to examine an ancient wootz blade [83]. The samples were taken from a sword dating from the 18th century. The global carbon content is 1.71 wt%. Electron micrographs of a sample examined at different levels of magnification (600 up to 100000) are presented in Figures 10.2.7, 10.2.8 and 10.3.3. Unlike carbides formed during Verhoeven's reconstitution, these carbides are sometimes faceted and irregular, arranged in less regular rows. In addition, their shape and distribution fit together, obviously suggesting that they were part of the same eutectic constituent (ledeburite). Another investigation of a different wootz blade points out a similar result, *i.e.* that coarsest particles could have the same origin in each cluster on account of the frequent similarity of orientation of the particles determined by Electron Backscattering Diffraction (EBSD), [93].

Precipitation of secondary carbides is illustrated in Figure 10.2.8, the very high proportion of precipitates corresponds to the high carbon content of this steel. Carbides formed by solidification of the liquid steel are said to be eutectic or primary and carbides formed by solid state precipitation are said to be secondary.

Similar microstructures were observed all along the transverse section of the blade. The poorly defined moire pattern indicates that the austenization was performed at a rather low temperature and that the number of heating/forging cycles was also insufficient.

10.3 Structure of the matrix

Austenite develops different microstructures according to the cooling rate as illustrated by the CCT curves in § 8.3. Genuine wootz forged following the traditional method possesses specific characteristics, essentially bound to the fact that it contains few alloying elements with a low content, together with a high carbon content. On quenching it may transform into plate martensite resulting in embrittlement of the blade. It seems that this microstructure was avoided by the ancient wootz steel blacksmiths. Let us quote a comment about Zschokke's findings [56]: *One of his conclusions, indicating that the composition of Damascus steel prevented the use of severe quenching, in order to avoid excessive brittleness, recalls a curious oriental tradition, concerning the air cooling of certain Damascus blades. As soon as forging was finished, while the blades were still red hot, they were given to a horseman, who galloped away furiously holding the blade in the air. This method of forced air cooling was probably more suitable than an abrupt quenching in cold water for these high carbon and phosphorus steel blades.* Mild cooling seems to be confirmed by the description of

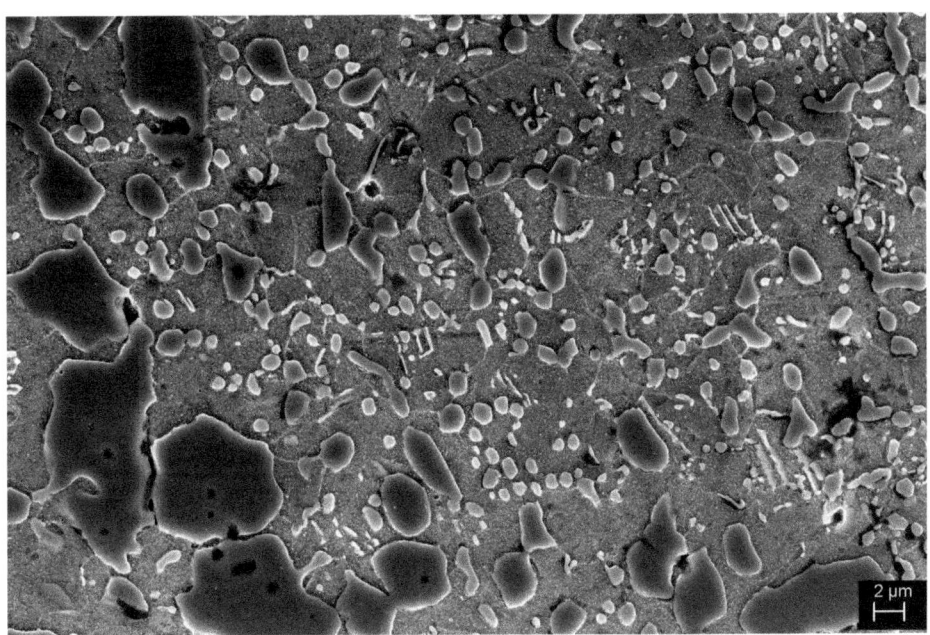

Figure 10.2.8 :
Type 1 carbides on the left side of the micrograph obviously fit together thus suggesting that they were part of a eutectic constituent (ledeburite).
Type 2 small carbides with various shapes and a size ranging between 0.4-3 µm precipitated in the austenitic matrix at temperatures above its transformation into pearlite, probably ranging between 900 and 600°C.
A few small lamellar pearlite islands indicate that pearlitic transformation was just begining and probably stopped by the martensitic transformation.
SEM electron micrographs, courtesy CMTC, INP Grenoble, Fr.

the Indian process whereby the red hot blades were plunged into the hollow trunk of a banana tree [9].

The different types of morphologies predictable are summarized in Table 10.3.1. In the frame are the ones corresponding to a soft quenching.

From the metallurgical point of view, bainite or tempered martensite microstructures correspond to excellent mechanical properties : high hardness and resistance associated with sufficient toughness. Modern bainitic steels, which are obtained by severely controlled heat treatments,

Table 10.3.1 : Transformation of austenite in high carbon steels as a function of increasing cooling rates

Divorced pearlite	Cells of spheroidized carbides
Pearlite	Cells formed by cooperative growth of alternate lamellae of ferrite and cementite.
Upper bainite	Plates which nucleate at the grain boundaries and grow in relation to specific crystallographic planes of austenite. In the case of high carbon steels the plates growing ahead are cementite instead of ferrite as is the case for low carbon steels. This bainite is sometimes named inverse bainite. The characteristic morphology is that of parallel plates at the border of the grain.
Lower bainite	Sheaves of ferritic laths with, inside the laths, small coherent metastable carbides precipitates along specific planes oriented with an angle 50-60° from the habit plane defined as the interface plane between austenite and bainitic lath as measured on a macroscopic scale.
Self tempered martensite	A structure rather similar to that of lower bainite.
Martensite	Laths in low carbon steels. Plates in high carbon steels with residual austenite resulting in brittlement of the as-quenched structure.

have been developed extensively. The steel hardness increases with the cooling rate, but it is not an univocal relationship; a given rate involves a certain hardness for each steel but conversely a given hardness does not allow us to deduce the thermal history of the sample because several paths are possible. The solid state transformations are strongly dependant upon the steel composition (Fig. 8.3.1, 8.3.1). Little variations involved in the ore supply, the charcoal used, the additions in the crucible and also the forging temperature (Fig. 8.3.2) are parameters which change significantly the steel behavior.

On a theoretical point of view, temperature chiefly influences the approach to phase equilibrium, the movement of interfaces and the growth mechanism of preciptates [70]. In the range of temperature approximately above the eutectoid temperature, the movement of the interface is a thermally activated process. At the front of the transformation matrix/new-phase the interface propagates readily in all directions. Below this temperature threshold, the interface prefers to move by lateral displacement of growth ledges. The microstructure then reveals facetted interfaces oriented along specific crystal planes.

Examination of wootz blades

Most often, the interest has been focused upon the distribution of carbides forming the moire pattern. The microstructure of the matrix was less investigated, assuming it was pearlitic. This seems to be inconsistent with the reputation of the exceptional mechanical properties of the blades and also with all the legends linked to the cooling processes. Anyway, it is very likely that the microstructure is not uniform all over the blade given that the thin parts cool faster than the core of the blade, even in the absence of coatings.

Verhoeven [94] gives in his experimental reconstitution a well identified example of a fine, globularized pearlite, said to be divorced, shown in Figure 10.2.3, corresponding to a rather low cooling rate, especially in the vicinity of the eutectoid temperature.

The micrograph of wootz observed by France-Lanord [95] (Figure 10.3.2) illustrates a different microstructure. The carbides are not as globular as common secondary carbides, they appear really elongated and as in dotted lines in some areas. It is attractive to imagine that these precipitates result from the partial spheroidization of plates or of rods. The proposed explanation is that they were formed in the carbon segregated zones as inverse bainitic plates which evolved during further moderate heating. The fact that these carbides are elongated or perfectly spherical hardly changes the macroscopic moire aspect because the important point is that they are distributed in alignments. In the case of high carbon steels, upper bainite (named also inverse bainite) can be confused with pearlite on account of the formation of parallel plates of cementite in both cases. To summarize, the interpretation assumed is inverse bainite in the segregated borders and tempered martensite in between.

The third example comes from the observation of our sword blade at very high magnification using a Scanning Electron Microscope - Field Emission Gun (SEMFEG) microscope. Figure 10.3.3 presents a magnified area taken from Figure 10.2.8-Top. The microstructure is similar on the whole transversal section of the blade. The coarser carbides, defined as type 2, are secondary carbides. In the matrix finer

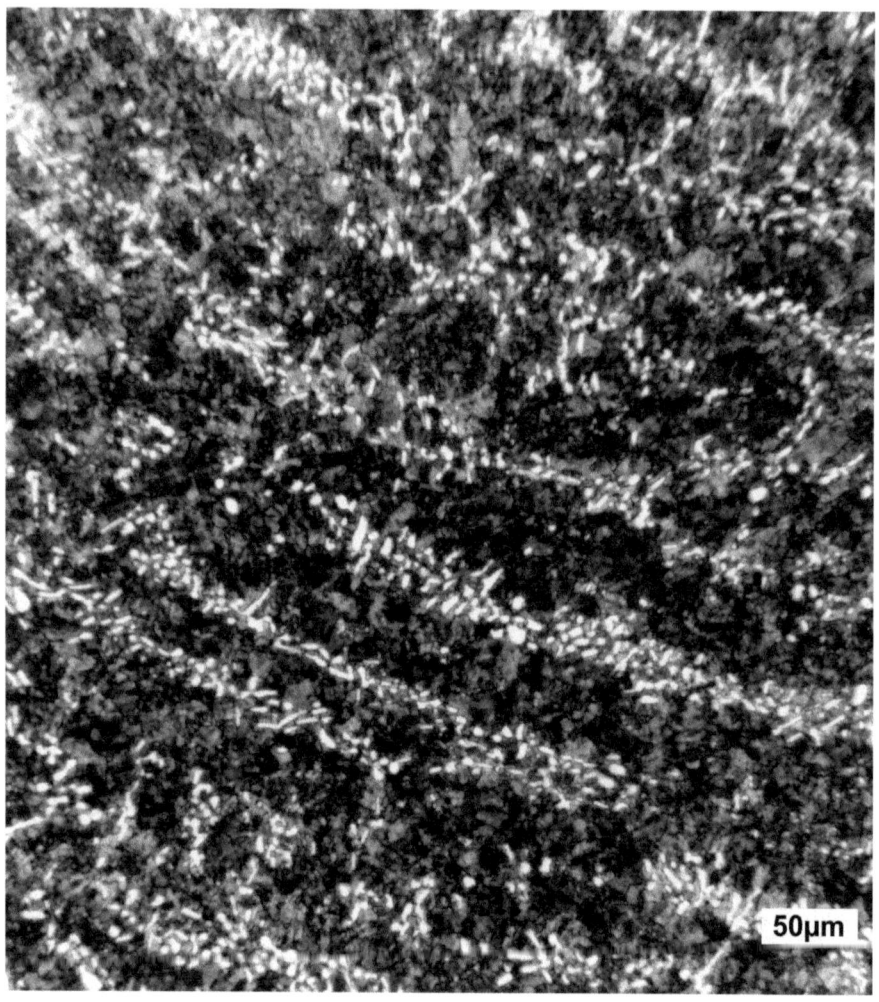

Figure 10.3.2 :
France-Lanord's optical micrograph of a wootz sample (Persia, 18th century [95]). Courtesy Laboratoire d'Archeologie des Métaux, CCSTIFM, Jarville, Fr.

carbides, labelled type 3, of a few tens of nanometers are regularly distributed. They are formed during the tempering treatment of a martensitic matrix at temperatures ranging between 200 and 600°C. The strong nital chemical etching highlights the protruding precipitates by dissolution of the matrix, they appear in clear contrast. In Figure 10.2.8-Bottom, a still finer precipitation can be detected in between white type 3 carbides. These type 4 precipitates of some nanometers

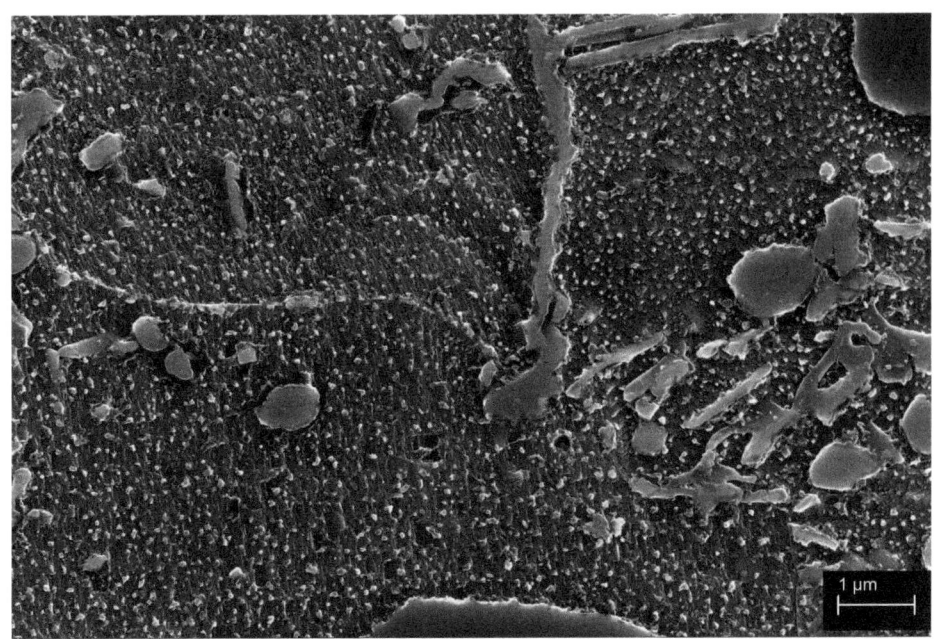

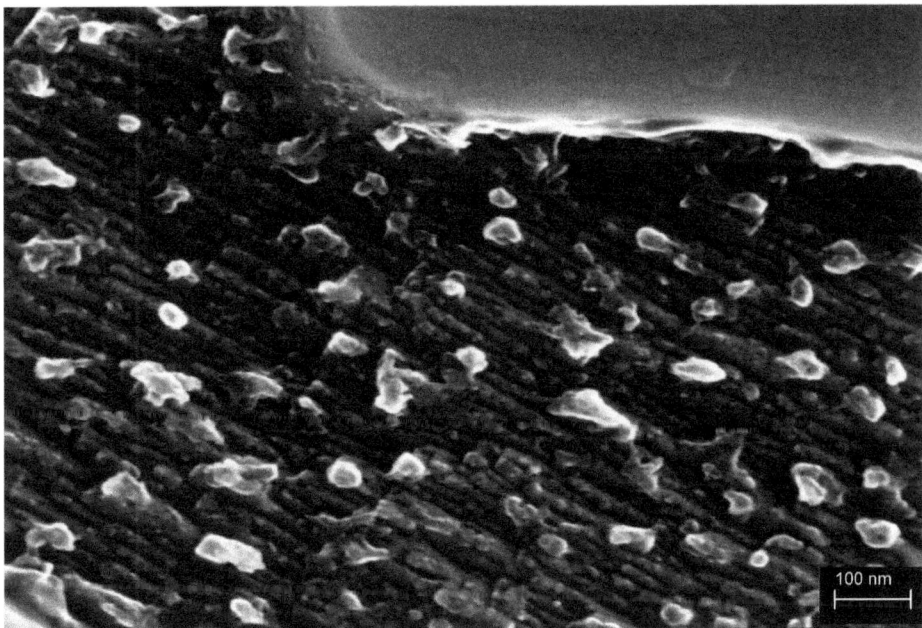

Figure 10.3.3 :
Electron micrographs of a wootz sample observed with very high levels of magnification. Courtesy CMTC, Fr.

are aligned according to certain directions. Considering the size, their temperature of formation is estimated to be very low, around 100°C, or possibly room temperature maintained for a long time during an aging process.

Another study [96] was conducted by a team of crystallographers on a sample of the same blade studied by Zschokke. A SEM micrograph illustrates a complex microstructure in which some areas exhibit the characteristics of tempered martensite and others those of a lower bainite. According to the authors this micrograph corresponds to the border of the blade while other parts are mainly fine grained pearlite [97-98].

To conclude, the wootz characteristic features are the result of transformations carried out during three distinct steps summarized in Table 10.3.4.

Table 10.3.4 : Transformations corresponding to each range of temperatures

Solidification	Segregation and formation of eutectic carbides
Forging range	Progressive formation of alignments, precipitation of secondary carbides mainly responsible for the cutting edge
Final cooling	Resistance and toughness of the matrix

Do alloys age ?

It should be recalled that the as-cooled structure is never a structure completely in an equilibrium state. The matrix of wootz is constituted with variable proportions of martensite and residual austenite, both supersaturated in carbon. An equilibrium state tends to be established thanks to the diffusion, primarily that of carbon, weak at ambient temperature which, nevertheless, can operate over very short distances on the nanometer scale.

The final equilibrium state of the binary system iron-carbon makes ferrite and graphite coexist alone. Does the evolution during several centuries tend to graphitize the alloy, *i.e.* to replace cementite by ferrite and graphite? The answer will doubtless be found by observations on a nanometric scale with the modern powerful means. No significant change seems to occur any on a micronic scale since relatively hard martensitic structures have been observed on samples dating from the first millennium BC.

"Modern wootz"

Blacksmiths succeeded in obtaining the structure in alignments starting from steels quite different in composition from the traditional wootz. Indeed, even if the available steels appear to be similar in composition to the ancient steels, modern industrial elaboration incorporates specific residual impurities, one of which is silicon. In the same way, certain steels proposed under the name *wootz* are traditional commercial grades, 100Cr6 or high speed steels which contain the usual additives of modern steels. Lastly, elements of addition were deliberately experimented by Perttula [99]. All these steels have in common with the *wootz* a high carbon content and they are forged in the as-solidified state. Inversely, they are easily transformed into martensite, even with a moderate cooling rate because of the presence of these additional elements. It is an essential difference with regard to ancient *wootz*. In addition, as they are high carbon steels, the martensitic transformation induces very high internal stress. They are brittle and must undergo modern specific stress relieving heat treatments (§ 8.4 and 8.8).

Forging high alloyed steels as stainless steels requires a large number of operations of heating then forging before obtaining the famous moire, a number much larger than in the case of the traditional wootz. But even so, some blacksmiths succeeded in restoring an aspect very close to that of ancient *wootz* (see Figure 7.7.4). It is significant for chromium contents beyond 12% and even more for certain alloyed steels comprising moreover molybdenum and nickel. These alloying elements inhibit the diffusion of carbon. The resulting decrease of the kinetics of the transformations explains that a greater number of operations are necessary.

The presence of alloying elements modifies other metallurgical features: carbides which precipitate are mixed carbides of iron and chromium instead of cementite, and the austenization treatments require a higher temperature (see §8.8).

As a conclusion to this chapter I wish to quote a proverbial sentence[3] from The Art of Poetry, Boileau (1674): *Cent fois sur le métier remettez votre ouvrage* translated from French by Soame (1892).

Gently make haste, of labor not afraid;
A hundred times consider what you've said;
Polish, repolish, every color lay,
And sometimes add, but oftener take away.

---

3. as it is used in blacksmiths' forums

11 *Alignments in medium carbon steels*

The watered structures do not appear only in carbon-rich steels. They are observed in crucible steels comprising a percentage of carbon lower than approximately 0.5 %. They are also present on the blades of Japanese swords and we found them in laminated steels. It is again an effect due to segregation. However, the models explained in the preceding chapter do not apply any more. It is necessary to consider another mechanism of formation in which it is the ferrite which is formed on cooling and not cementite. Metallurgists would remark: this mechanism is not new, it is rather similar to that of the formation of the bands in steels often called banding.

11.1 A well-known phenomenon

At the end of the thermomechanical treatment of steels, darker bands, stretched out in the direction of rolling are visible at the surface, they are due to selective oxidation. The cause of this banding is a remnant of an alloying element or carbon segregation. Modern rolling installations that are more powerful and faster make it possible to avoid this phenomenon.

Figure 11.1.1-Top shows an optical micrograph of a typical low carbon steel. The dark bands are pearlite and the clear ones ferrite. Figure 11.1.1-Bottom shows an electron micrograph of a martensitic low alloyed steel. The large clear bands are martensite, while the dark bands consist in grains with a small ferritic core indicated by the arrow surrounded by pearlite. The mechanisms involved are similar in both cases.

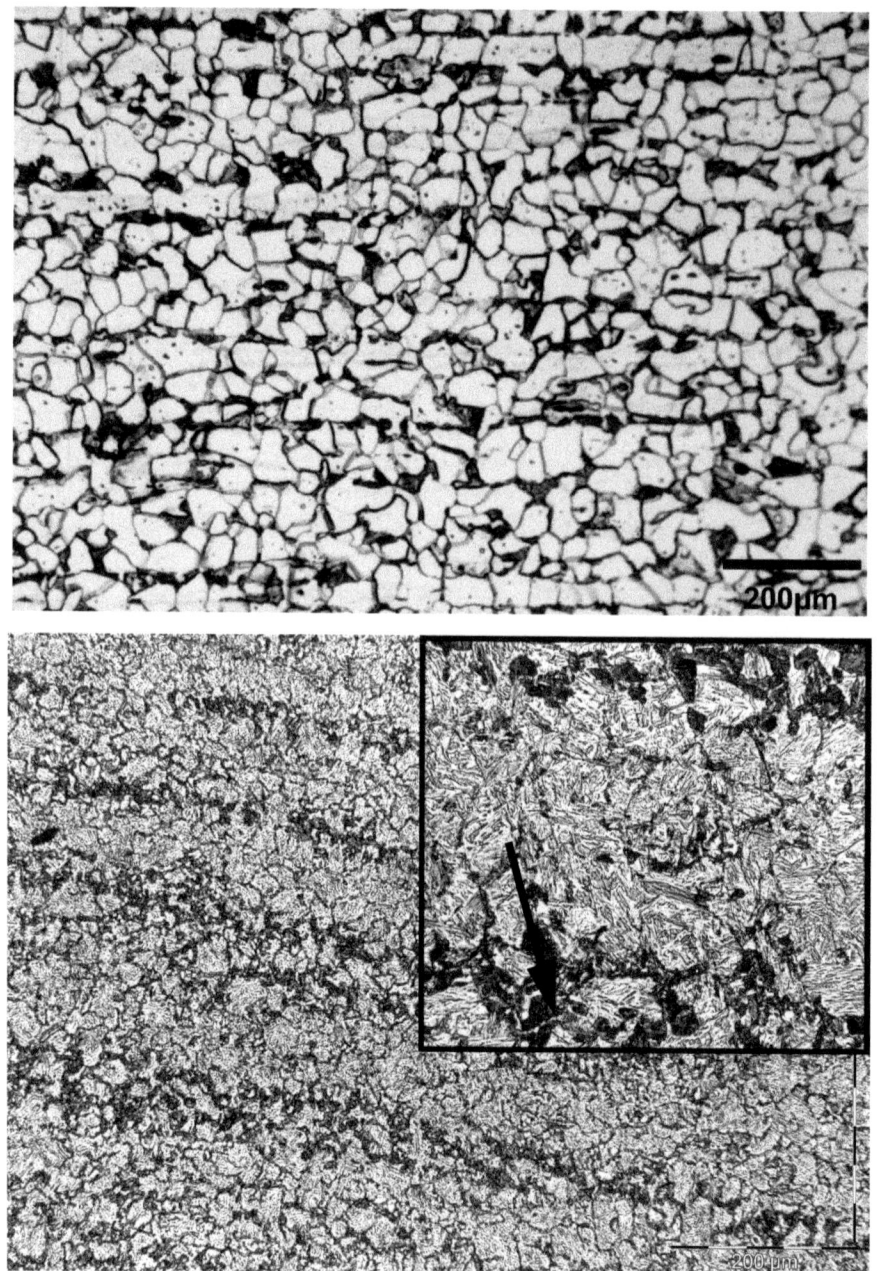

Figure 11.1.1 :
Micrographs of as-rolled steels : Top : Low carbon ferritic steel; Bottom : martensitic steel, the arrow shows the ferrritic nucleus.
Courtesy Grenoble INP, Fr.

The existence of residual segregation has been evidenced, segregation resulting from dendritic solidification which was not completely blurred during the austenization. Rolling deforms the austenite matrix, lengthens the grains and also spreads the associated segregation in the direction of rolling, resulting in a distribution of the segregated alloying elements along bands. This effect was studied in the case of steels on an industrial scale and manganese was pointed out as one of the highly responsible elements [100-102]. There is also another possible effect related to the recrystallization grain size but the problem does not arise for a strongly hammered steel.

During the cooling process, austenite is transformed into ferrite at a temperature determined by the local composition. According to the phase diagrams, these temperatures can differ by several tens of degrees between zones with different compositions (Figure 11.1.2). The presence of manganese, chromium and carbon lowers the temperature of ferrite formation and this gap is responsible for the banding.

The major influence of the cooling rate

Let us examine step by step the succession of transformations after solidification. In the range of temperature 1400-900°C (indicated as austenite II on diagram 11.1.2) few events occur, except a partial homogenization of carbon and a very limited homogenization of the alloying elements. Below 900°C approximately in zone III the precipitation of ferrite begins. The first precipitates nucleate in the more depleted zones of alloy elements and carbon, contrary to the precipitation of carbides in high carbon steels (Figure 11.1.3). The ferritic nuclei coarsen and progress while the elements rejected by the ferrite, mainly carbon, drain in front of them.

It is the carbon diffusion which controls the rate of advance of the growth front of ferritic grains. If the cooling is slow enough, the growth remains compatible with the diffusion rate. If cooling is too fast, the diffusion is not effective any more. Everything stops, the nucleation of ferrite takes place in many sites in an anarchistic way. Taking into account the small variations in composition between the cores and the borders of dendrites, the shifts of temperature are weak, about 20°C, which leaves a little interval to avoid generalized nucleation. The mechanism is validated by a systematic study of the micro-

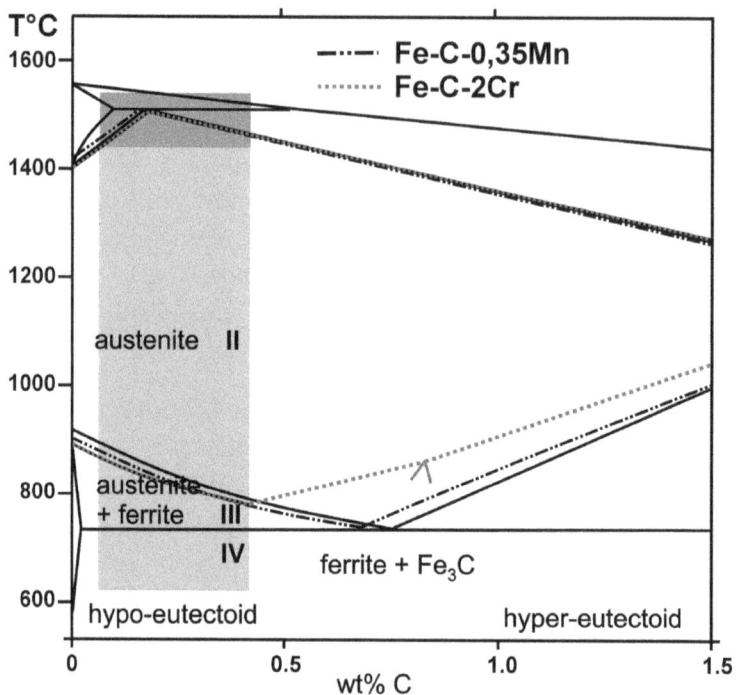

Figure 11.1.2 :
Calculated phase diagram Fe-Fe₃C with superimposed isopleths from ternary systems with respectively 0.35% Mn and 2% Cr.
The grey band indicates the composition range in which the banding is likely to occur.
Document INP Grenoble, Fr.

structure of a steel according to the cooling rate (Figure 11.1.4), [103]. Only low cooling rates generate a structure in bands. High cooling rates lead to uniformly distributed bainite or/and martensite.

Let us return to the case of Figure 11.1.1 B shown as an example of bands in a martensitic steel. The industrial alloy contains elements such as chromium, which inhibit the diffusion of carbon and slow down the progression of the ferritic interface. The transformation remains at the first stage of diagram 11.1.3, a small ferrite grain progresses sweeping a diffusion layer readily transformed into pearlite while the matrix transforms into martensite.

Figure 11.1.3 :
Formation of strips :
I) in a segregated matrix ferrite nucleates in the more carbon depleted zones (clear contrast) ;

II) at the interface between austenite and ferrite, the austenite transforms into ferrite which progresses towards the border of the dendrite. Rejected carbon accumulates in a narrow diffusion layer and diffuses ahead the interface in the untransformed matrix.

III) ferritic grains coarsen. Interfacial grain boundary between ferrite and austenite is a highway for carbon diffusion, thus allows draining of carbon and merging of the ferritic grains.

IV) pearlite forms when the carbon content in austenite reaches 0.8% and the temperature is below the eutectoid temperature.

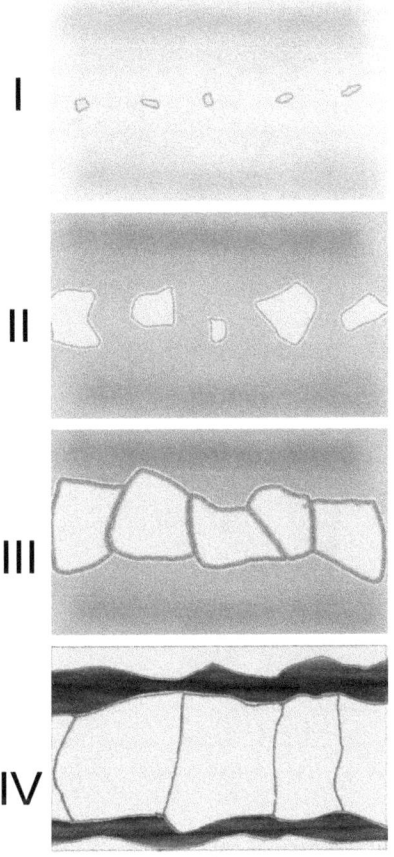

11.2 Occurrence of banding in ancient steels

In welded steels

The knives found at Paladru were made with laminated steels hot worked during multiple cycles of folding and forging/welding. The blade shown in Figure 11.2.1 corresponds to a very low carbon steel, nearly pure iron. The dark strips are fine layers of iron oxide entraped by folding. During transformation of austenite into ferrite the growth of the grain is only stopped by these layers. This generates a typical *bamboo* structure. An ESBD investigation was made in which the color of the grains corresponds to a specific orientation. On the micrograph, each grain

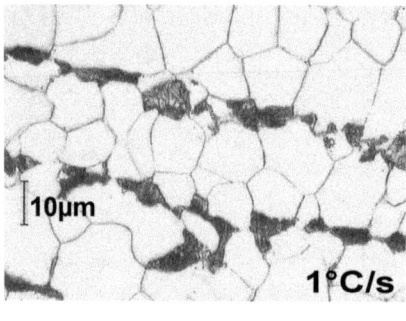
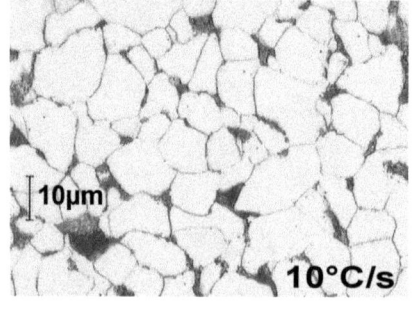
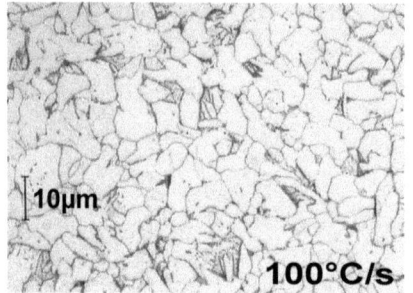

Figure 11.1.4 :
Optical micrographs of a steel Fe-0.11C-0.45Mn for increasing cooling rates. Nital etching reveals ferrite in clear contrast and pearlitic cells in dark contrast.
Courtesy Universities of New South Wales and of Wollongong, Australia [103].

appears with a shade different from that of its neighbors (which is more evident on the corresponding colored image), this means that they nucleated independently.

The second example of banded structure shown in Figure 11.2.2 is another blade of the Paladru knives in which strips of pearlite and ferrite alternate (also presented in Figures 5.2.2 and 5.2.3). The high number of welding tracks evidences that the steel has been hot worked with many cycles. A segregation gradient of carbon can be formed when the bar is heated, the superficial zone can be carburized or decarburized according to the adjustment of the hearth (Fig.9.1.1). This zone then integrated into the matrix in which the uncomplete homogenization is likely to initiate ferrite nucleation of the first bands. The micrographs shown in Figure 11.2.3 are intended to highlight the oxide precipitates which underline the welds. The more ancient welds, which have been repeatedly heated, appear globurized for example the ones shown in Figure 9.1.1. Comparison of top and bottom images evidences that the alternation of pearlite/ferrite is spaced quite independently of the many welds.

Only the shape and spacing of the oxide inclusions are considered in this discussion because they characterize the forging process. However, analyses of the slag chemical composition may be very helpful in determining the origin of the ore, or informing about the bloomery process [17-104].

Figure 11.2.1 : Bamboo microstructure. Cross section of a knife from Paladru and EDBD image of a magnified area. Courtesy INP Grenoble, Fr.

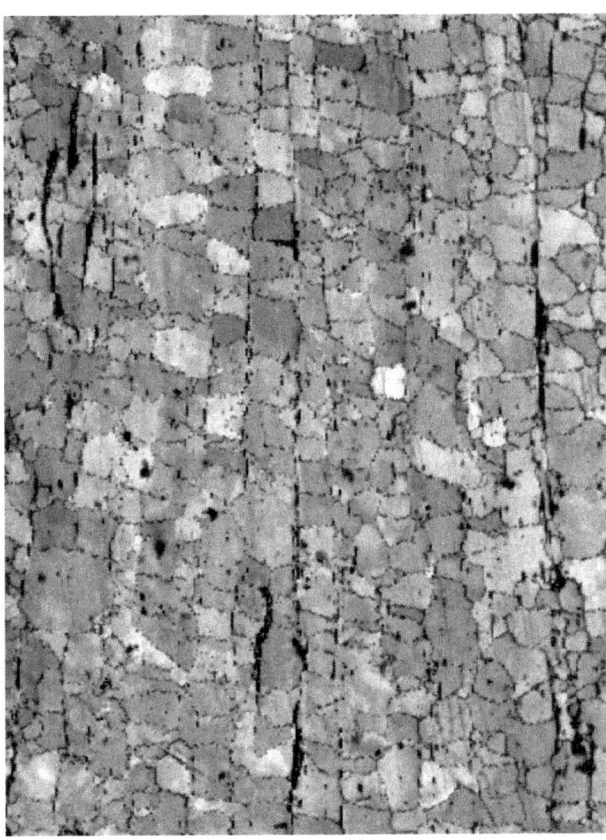

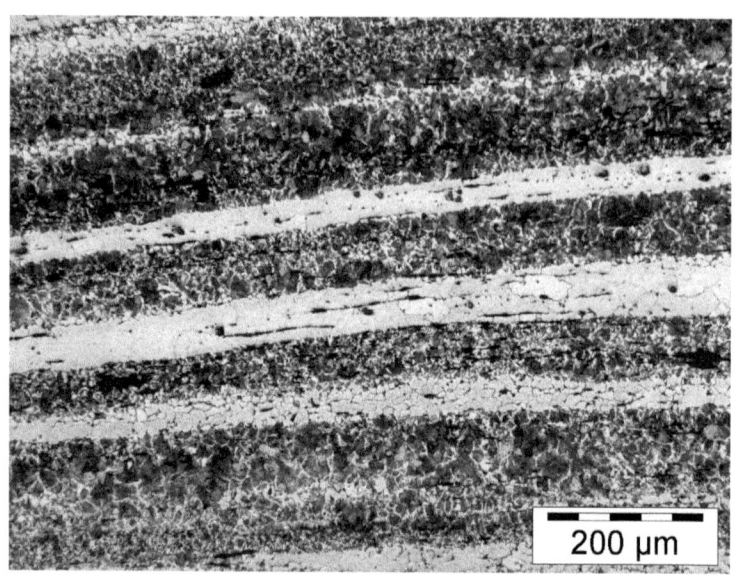

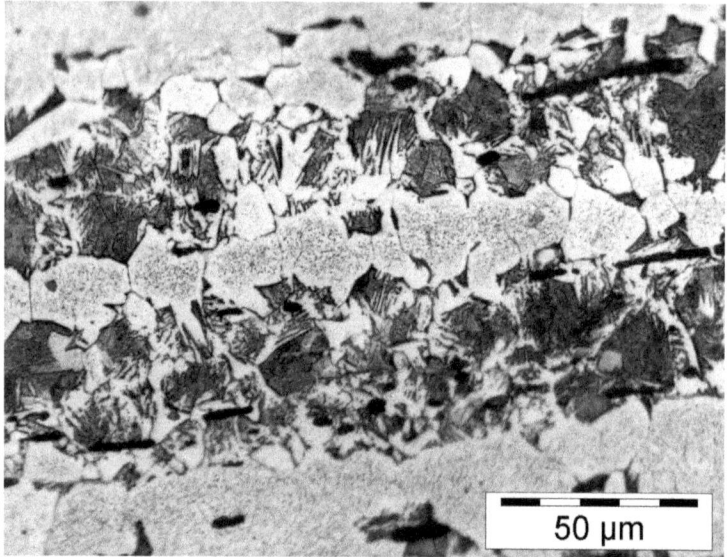

Figure 11.2.2 :
Optical micrograph and magnification observed on the cross section of the blade showing alternated bands of ferrite and pearlite. Sample of a Paladru knife shown in Fig. 5.2.2 of CCSTIFM, Jarville, Fr. ;
Document INP Grenoble, Fr.

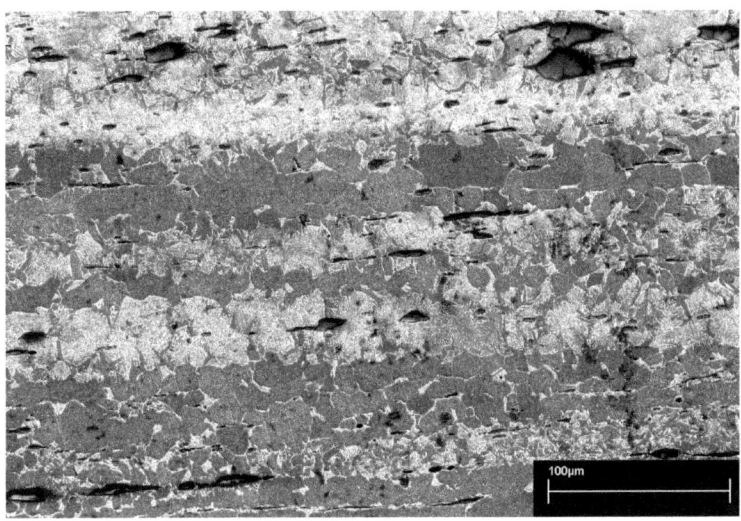

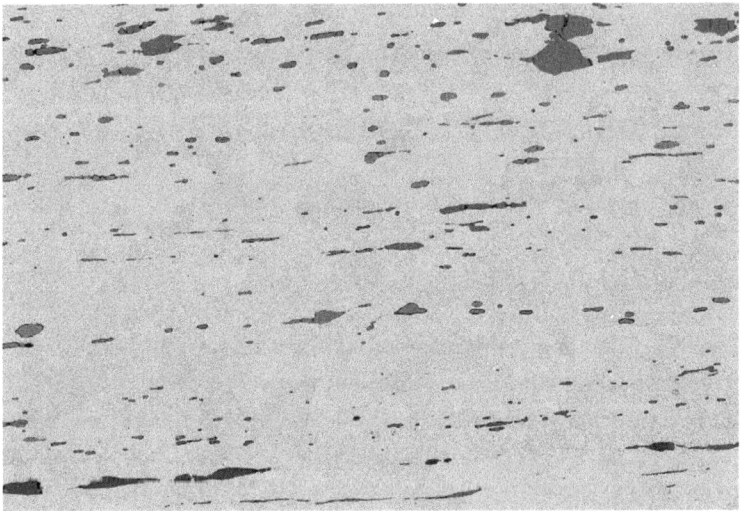

Figure 11.2.3 :
Electron micrographs of the same sample as in Fig. 11.2.2 showing:
top a secondary electron image of alternated bands and
bottom a backscattered electron image of the same area in which the entrapped oxides appear in dark contrast.
Document INP Grenoble, Fr.

The decarburizing forging of the katana blades

The starting step of the forging of a katana is hot working steel with approximately 1.2 % carbon content (§ 6.7). The steel undergoes a succession of operations during which it looses carbon superficially. Many cycles of heating/forge-welding must be carried out to impoverish steel until a global composition of about 0.4% C is acheived *i.e.* that of a hypo-eutectoid steel. The final microstructure is an alternation of more or less carburized layers, again a pearlite/ferrite structure results in a visible pattern on the surface of the blade, it is the *ji-hada*.

The *ji-hada* can have different patterns depending on how the steel is worked, it can be folded in one direction lengthwise only, or, it can be folded alternately lengthwise and crosswise. The way the stacks of steel are hammered is important; changing the strength of the hammer blows, the shape of the hammer head or the temperature of hot working induces variations in the pattern.

The visible patterns thus obtained are named after Kapp *et al.* [62-63]:
Masame-hada a straight pattern
Mokume-hada a pattern in which small ovals or circular patterns are seen similar to a fine burl grain seen in wood
Itame-hada a regular weavy pattern resembling a wood grain in which irregular ovals appear.
Ayasugi-hada a regular weavy pattern.

The forged metal having been folded thirteen times or more, becomes a kind of metallic pastry dough consisting in 2^{13} layers *i.e.* 8192 per inch of steel resulting in a spacing of about $3\,\mu m$. The arrangement of the layers of pearlitic cells is much coarser since it is visible with the naked eye. It means that the formation of bands is independent of the weld tracks, and remains only a diffusion controlled mechanism.

Good polishing is very difficult because ferrite is ductile and pearlite is also rather soft; hard polishing pastes like diamond pastes must be avoided since hard particles incrust in the soft matrix. An awkward polishing would cause a spreading out of the layers the ones on the others. The polisher, writes Kapp, *with an arsenal of small stones of ever increasing fineness, is able to make visible and attractive all the details of the smith's metalwork.*

Soft damask

All the ancient crucible steels were not carbon-rich steels. However blades forged with low-carbon steels (content lower than 0.5%) can develop the moire or stripey patterns qualifying them to be *soft damask* [32], (§6.6). An example of such a blade is given by Verhoeven [59]. This blade though not considered as *true wootz Damascus steel* could be produced by crucible melting.

The interpretation used for the formation of alignments in the carbon rich wootz is not usable anymore since cementite precipitation is not possible in hypo-eutectoid alloys. The pattern, as far as there is one, can be produced only by the presence of pearlitic cells arranged in alignments. This is the mehanism of banding explained in detail about the knife from Paladru, and like in Verhoeven's model there remains the paramount role of segregation.

This damask is less contrasted than in the case of wootz even with an appropriate chemical etching. On account of a difficult polishing genuine blades are often poorly highlighted. In addition, since this steel may rust, the blades presented in exhibitions or Museums are sometimes protected by a varnish and then, the moire pattern of wonderful blades is completely blurred by a greyish varnish. If there is any lost know-how about wootz, it is the art of achieving a good polish!

11.3 The contribution of structural metallurgy

Practice of forging in order to generate bands

To summarize, banded microstructure can be initiated in the solid solution of austenite during forging. Precipitation nucleates in zones enriched or impoverished in elements, whatever the origin of the heterogeneity: either segregation formed during solidification (Verhoeven model), or segregation formed during the cooling process by grain boundary precipitation (Wadsworth model). The banding effect is enhanced at each cycle provided that the warming-up time is short enough and that the cooling rate is relatively slow in the forging range of temperatures between 900 and 750°C approximately (see field III in Figure 11.1.2). Several tens of cycles are necessary to obtain a well defined moire pattern.

The steel should be heated, neither too high to remain above the incipient melting temperature, nor during a long time in order to avoid erasing the segregations. Moreover, forging at high temperature for a too long time ends up transforming cementite into graphite distributed in alignments in the place of ancient carbides, falling the mechanical resistance.

Feeling nostalgia for the time when wootz was a mystery !

Wootz is not really any more a mystery, it is simply a complex problem because it is clear that modern metallurgical knowledge makes it possible to explain the formation of alignments. There remains still work to be done to analyze mainly the structure of the matrix and compare samples of various periods and various origins. Very numerous parameters are to be taken into account to enter all the complexity of the problem. The thermal parameters are essential, unfortunately these latter are very poorly defined during forging in the traditional way, consequently it is not possible to compute a miraculous formula for each grade. The skill, the experience and the patience of the blacksmith remain essential qualities for the success of original works.

Expertise in crucible steel

A banded microstructure does not evidence that the steel was *crucible steel,* inversely crucible steel produces a banded structure only when it is forged in the conditions defined above. Forging wootz consisted, during the second millenium, in hammering and stretching the cake to give the shape of a blade without repeated folding. The process does not generate a web of oxides. Then, can this microstructural feature be considered as specific to crucible steel? To support this discussion a summary of the different possibilities is made in Table11.3.1.

This discussion was suggested by recent findings about Viking age swords. Williams [105-106] studied swords of the Wallace collection dating from this period. Microanalysis and microscopic examinations revealed that the specimens with the Ulfberht inscription, which were reputed to be hard and tough, have a rather high carbon content. The author reports [107]: *the swords were made of imperfectly melted steel - consisting of a mixture of iron and carbonaceous materials heated together to give high-carbon steel. NPL's results match descriptions of ancient sword making in Herat (now in Afghanistan) described by ninth century Arab philosopher and writer Al-Kindi..*

Table 11.3.1 : Distribution of inclusions according to the process

Wrought smelted steels	Most of the slag inclusions, initially rather coarse, are expelled from the lump by hot working. Fine oxide inclusions remain, even when the steel is hammered for a long time, see Figures 2.2.4, 2.2.5, 5.2.2, and 5.2.3.
Melted steels	Melting steel induces partitioning of slag which floats above the liquid steel and drag inclusions, leaving only a low proportion of inclusions. This is the case of wootz cakes prepared by fusion in a crucible. Modern steels are also produced by fusion with procedures allowing a high level of cleanliness.
Forge-welding	During hot working the free surface of a part is oxidised and when two parts are forge-welded, the oxides are entrapped into the weld. Cycles of folding and forge-welding induces the formation of a regular distribution of small oxide inclusions. An example is shown in Figure 9.1.1.
High temperature welding	When the welding process is performed at high temperatures close to the temperature of the begining of fusion of one grade, the welded zone extends in a rather large band of about 50µm in which tiny oxides are scattered, see Figures 7.2.4, 9.1.3.

About 300 Ulfberht swords have been found, they were produced over a period 850-1100 AD. It is well established that during this period, the Vikings exploited a trade route from the Baltic to Persia down the Volga and across the Caspian Sea. The use of this route declined after the collapse of the Sassanid Persian empire (819-999). The manufacture of Ulfberht swords ceases presumably because crucible steel was no longer available.

Then, history supports the assumption that crucible steel was used to manufacture Ulfberht swords. A sound evidence is difficult to obtain only from microstructural observations. In addition, a criticism was made about the sampling of the blades examined; most of the samples cover only the edge of the blades [108]. Many other questions remain unsolved. Where and how were the blades forged? Was crucible steel imported in the shape of cakes or bars (something similar to salmon of swords). Did blacksmiths know the Persian swordsmithing methods or did they use their own methods?

To conclude

I would like to underline the deep evolution of the metallurgy of steels, evolution which is regrettably not visible to the general public. Steel appears as an immutable material as far as it can be completely recycled, and its metallurgy is considered as an achievement, part of world heritage. The fact that one produces steels much more resistant to shocks, to stresses, to fatigue, to high or very low temperatures and to corrosion is not spectacular. Admittedly, the consumer appreciates that the body of cars rusts less quickly, or that the modern ships are more reliable than the Titanic.

However progress from the last 50 years was considerable. This is due mainly to the invention of electron microscopy which, just as the invention of optical microscopy in the 17th century, allowed a great jump in the characterization of the structure of steels. The optimization of a steel grade is not made any more by the simple examination of abacuses for properties of employment. On the other hand, modern steels require a precise adjustment of composition accompanied by rigorous procedures of elaboration and heat treatments in order to develop the best level of performances.

The availability of powerful instruments makes possible the observation of metal on an increasingly fine scale down to the local arrangement of the atoms. However steels are a very heterogeneous material in the majority of the cases, and especially high carbon steels. It is thus imperative (I insist on that point) to be attentive to the structure of metal at various scales from the dendritic grain to the nanometric precipitates. This vision of multiple scales makes even more necessary to understand the mechanisms concerned. As a consequence modern structural metallurgy becomes a powerful tool for expertise.

12 *References*

12.1 List of references

[1], O.D. Sherby and J. Wadsworth, "Ancient Blacksmiths, the Iron Age, Damascus Steels, and Modern Metallurgy", *Thermec 2000* Las Vegas and *Journal of Materials Processing Technology,* **117**, Issue 3, (2001), 347–353.

[2], V.F. Buchwald, *Handbook of iron meteorites*, University of Californie Press, Los Angeles CA, USA (1975).

[3], *Les alliages de fer et de nickel*, Ed. G. Béranger, F. Duffaut, J. Morlet, J.F. Tiers, Pub. Lavoisier, Paris (1996).

[4], P. Z. Budka, "Meteorites as Specimens for Microgravity Research", *Metall. Trans.* **19A** (1988), 1919-23.

[5], *Le livre de l'acier*, (1994) and *The book of steels*, Ed. G. Béranger, G. Henry, G. Sanz, Pub. Techniques et documentation, Lavoisier, Paris (1996).

[6], D. Johnson, J. Tyldesley, T. Lowe, P. J. Withers, M. M. Grady, "Analysis of a prehistoric Egyptian iron bead with implications for the use and perception of meteorite iron in ancient Egypt", *Meteoritics & Planetary Science*, **48**, **6**, (2013), 997–1006.

[7], E. Salin, "La civilisation mérovingienne" 3ème partie: *Les techniques,* Edité avec le concours du CNRS, Picard, Paris (1957).

[8], J. Ypey, "Au sujet des armes avec damas soudé en europe", *Archéologie médiévale,* Tome **XI**, Centre de Recherches Archéologiques Médiévales, Caen Fr. (1981), 147-165.

[9], B. Prakash, "Some Aspects of Process Control of Ancient Iron making", *Paléométallurgie du fer et cultures*, Ed. P. Benoit et P. Fluzin, Pub. Vulcain Belfort, Fr, (1995), 33-40

[10], S. Srinivasan and S. Ranganathan, "Wootz steel: an advanced material of the ancient world", *Iron and steel Heritage of India*, Ed. S. Ranganathan, Jamshedpur: Indian Institute of Materials and Tata Steel (1997).

[11], S. Srinivasan and S. Ranganathan, *India's legendary "Wootz" Steel: an Advanced Material of the Ancient World*, National Institute of Advanced Studies, Bangalore, and Indian Institute of Science, Bangalore, (2004).

[12], A. Feuerbach, *Crucible Steel in Central Asia: Production, Use, and Origins*, PHD University of London, (1998).

[13], A. Feuerbach, "Damascus Steel and Crucible Steel in Central Asia", *American Society of Arms Collectors Bulletin,* **82**, 33-42

[14], A. Feuerbach, "The glitter of the sword: the fabrication of the legendary damascus steel blades", *Minerva*, July/August 2002, 45-48.

[15], A. Feuerbach, "Crucible Damascus Steel: a Fascination for Almost 2,000 Years", *JOM*, May 2006, 48-50

[16], L.S. Figiel, *On the Damascus Steel*, Atlanta Arts Press, Atlanta (1991).

[17], J. Le Coze, "Purification of Iron and Steels a Continuous Effort from 2000BC to AD2000", *Materials Transactions, JIM*, **41** N°1 (2000), 219-232.

[18], J.P. Mohen, *Métallurgie préhistorique. Introduction à la paléométallurgie*, Masson, Paris (1990).

[19], H. Rubin, T. Ko, "Ancient Iron and Steel Technology in China", *Paléométallurgie du fer et cultures*, Ed. P. Benoit et P Fluzin, Pub. Vulcain Belfort, Fr. (1995), 111-117.

[20], D.B. Wagner, *Iron and Steel in Ancient China*, Ed. Brill, Leiden (1993).

[21], D.B. Wagner, "The earliest use of iron in China", *Metals in antiquity*, ed. by S. M. M. Young, A. Mark Pollard, Paul Budd and Robert A. Ixer, Oxford: Archaeopress, (1999), 1-9.

[22], G. Hongye and H. Jueming "Research on Han Wei Spheroidal-graphite Cast Iron", *Foundry Trade Journal Intern.* March 1983, 89-94.

[23], Q. Hancheng, and P. Xiaodong, "Study of the Formation of Spheroidal Graphite in Ancient Cast Iron in China" *AFS Transactions*, **99** (1991), 415-420.

[24], D. Grébénart, "Reassessment of the Evidence for Early Metallurgy in Niger, West-Africa", *J of Archæological Science*, USA **V15** (1988), 367-394.

[25], R.J. Forbes, *Studies in Ancient Technology* **IX**, Ed. E.J. Brill, Leiden (1964).

[26], C.S. Smith, *A History of Metallography*, The Univ. of Chicago Press, Chicago & London (1965).

[27], R.F. Tylecote, *The early history of metallurgy in Europe*, Longman archeology series, Pub. Longman INC. New York USA, (1987).

[28], R. Pleiner, "Les débuts de la métallurgie du fer chez les Celtes", *Les Princes Celtes et la Méditerranée*, La documentation française, Paris (1988), 179-185

[29], J.Y. Andrieux, *Les travailleurs du fer*, Découvertes Gallimard, Sciences et Techniques, Ed. Gallimard (1991).

[30], J.E. Rehder, "Iron versus bronze for Edged Tools and Weapons. A Metallurgical View", *JOM,* August 1992, 42-46.

[31], M. Eliade, *Forgerons et alchimistes*, Champs Flammarion, Paris, (1977).

[32], M. Sachse, *Damascus Steel*, Ed. Stahleisen, Düsseldorf, Germany (1994).

[33], P.R. Schmidt, *Historical Archaeology, A Structural Approach in an African Culture*, Greenwood Press, Westport, Conn. USA (1978).

[34], C. Éluère, *L'Europe des Celtes*, Découvertes Gallimard, (1992).

[35], L. Bonnamour et A. Dumont, "Les armes romaines de la Saône: état des découvertes et données récentes des fouilles", *JRMES* **5**, (1994), 141-154

[36], H. Meyer-Roudet, *A la recherche du métal perdu. Nouvelles technologies dans la restauration des métaux archéologiques.* ouvrage collectif, Ed Errance, (1999).

[37], A. France-Lanord, "Le fer en Iran au premier millénaire avant Jésus-Christ", *Revue d'histoire des mines et de la métallurgie*, **1**, n° 1, 75-126.

[38], C. Forrières, "Etude par microscopie électronique de structure de trempe d'une lame d'épée du Luristan", *Revue d'Archéométrie*, **11**, (1987), 17-29.

[39], R. Pleiner, *"The Celtic swords"*, Oxford University Press, Clarenton, USA, (1993).

[40] P. Fluzin et L. Uran, G. Béranger, C. Coddet, "Structures et mise en forme d'armes gauloises", *Rev. archéologique de Picardie*, 1, (1983), 181-194.

[41], A. France-Lanord, "Les techniques métallurgiques appliquées à l'archéologie", *Rev. Mét.* **XLIX**, 6 (1952), 411-422.

[42], M. Ehretsmann, "L'épée carolingienne de Strasbourg", Cahiers Alsaciens, *Art et Histoire*, **31**, (1988) 85-94.

[43], J. Renaud, *"Les Vikings en France"*, Editions Ouest-France, (2000).

[44], B. Solberg, "Weapon export from the Continent to the Nordic Countries in the Carolingian Period", *7 Studien zur Sachsenforschung*, Bergen, (1991) 241-259.

[45], I. Peirce, *"Swords of the Viking Age"*, The Boydel Press, Woodbridge, UK (2002).

[46], M. Byam, *Armes et armures*, Ed Gallimard, Les yeux de la découverte, Paris (2003).

[47], R.N. Balsiger, E.J. Kläy, *Bei schah emir und khan*, Meier Verlag, Schaffhausen, (1992).

[48], Manouchehr M. Khorasani, *An Introduction to the Persian Sword Shamshir,* Internet, (2006)

[49], Manouchehr M. Khorasani, *Lexicon of arms and armor from Iran a study of symbols and terminology,* Legat Verlag (2010)

[50], Jean-Jacques Perret, Maître coutelier de Paris, " *Les arts du coutelier et du chirurgien*" M DCC LXXI (1771)

[51], J.F. Clouet, "Instruction sur la fabrication des lames figurées dites damassées", *Mémoire publié dans le Journal des Mines*, **5** n°90, Ventose an XII (1803).

[52], H. Landrin, *Manuel du coutelier, traité théorique et pratique de l'art de faire tous les ouvrages de coutellerie*, Paris (1835).

[53] G.J. Mac Donald, "Spanish textile and clothing nomemclature in -an, -i, and -in", *Hispanic Review*, Vol. 44, **1**, (Winter 1976), 57-78.

[54], M. Collardelle et E. Verdel, "Le mobilier métallique : typologie", *Les habitats du lac de Paladru (Isère) dans leur environnement*, dAF **40**, Editions de la maison des sciences de l'homme, Paris (1993), 204-219.

[55], C. Forrières, P. Merluzzo et A. Ploquin, "La métallurgie du fer", *Les habitats du lac de Paladru (Isère) dans leur environnement*, dAF **40**, Editions de la maison des sciences de l'homme, Paris (1993), 220-237.

[56], B. Zschokke, "Du Damassé et des lames de damas", *Rev. Mét. Part I*, **21**, (1924), 635-669.

[57] C. Panseri, "L'acciaio di Damasco nelle legenda e nelle realtà", *ARMI ANTICHE*, Bollettino del Accademia di S. Marciano, Torino (1962).

[58], J. Wadworth and O. Sherby, "On the Bulat-Damascus Steels Revisited" *Progress Materials Science,* **25**, (1980), 35-68.

[59], J.D. Verhoeven, A.H. Pendray, and W.E. Dauksch, "The Key Role of Impurities in Ancient Damascus Steel Blades" *JOM,* Sept. 1998, 58-64.

[60], J.D. Verhoeven, A.H. Pendray, and E.D. Gibson, "Wootz Damascus Steel Blades", *Mater. Char.* **37** (1996), 9-22.

[61], R Zeller, und Ernst Rohrer, "Orientalische Sammlung Henri Moser-Charlottenfels. *Beschreibender Katalog der Waffensammlung*, Bern 1955

[62], L. Kapp, H. Kapp and Y. Yoshihara, *The Craft of the Japanese Sword*, Ed. Kodansha Int. (1987), Tokyo, Japan.

[63], L. Kapp, H. Kapp and Y. Yoshihara, *Modern Japanese Swords and Swordsmiths*, Ed. Kodansha Int. Tokyo, (2002).

[64], E.C. Bain, "Nippon-tô, an introduction to old swords of Japan", *J.I.S.I.* April 1962, 265-282.

[65], H. Tanimura, "Development of the Japanese Sword", *J. of metals,* Feb 1980, 63-73.

[66], Tatsuo Inoue, "Japanese sword", *Materials Science Research International*, **2** N°4 (1997), 193-203.

[67], Tatsuo Inoue, "Japanese swords in comparison with others", in *Progress in Mechanical Behaviour of Materials* Proceedings ICM8 **II** *Materials Properties* Ed. F. Ellyin, J.M. Provan (1999).

[68], *Acier DAMAS. L'acier corroyé occidental.* Ed. Musée de la coutellerie de Thiers, Maison des couteliers, (1999) CDrom.

[69], J.D. Verhoeven, and H.F Clark, "Carbon Diffusion Berween the Layers in Modern Pattern-Welded Damascus blades", **41** (1998), 183-191.

[70], M. Durand-Charre, *Microstructure of steels and cast irons*, Ed. Springer, (2004), Berlin.

[71], M. Pohl, "Gefüge und Eigenschaften von Damast- und Duplexstälen", *Fortschritte in der Metallographie*, Ed. P. Portella, 2002, 254-261.

[72], Billgren et al. United States Patent; Patent Number 5,815,790; Date Sept. 29, 1998.

[73], P. and M. Billgren, *Damasteel handbook*, Damasteel AB (1999)

[74], G. Obach, *Replication of Wootz "Damascus" Type Steel,* Report Laurentian University, Canada, Internet, (2003)

[75], Manouchehr M. Khorasani, "Reviving the ancient art of making persian crucible steel for bladed weaponry", *Journal of Asian Martial Arts*, **17**, **1**, 54, (2008).

[76], G. Krauss, *Principles of heat treatments of steels* , Ed. ASM USA (1980).

[77] H.K.D.H. Bhadeshia, R.W.K. Honeycombe, *Steels: Microstructure and Properties*, Butterworth-Heinemann Ltd; 3rd Revised edition edition (2006).

[78], J.D. Verhoeven, *Steel Metallurgy for the Non-Metallurgist*, ASM International, (2011).

[79], G. Béranger et D. Henriet, "Coloration des aciers inoxydables", *Techniques de l'Ingénieur, traité des matériaux métalliques*, M 1 572, 1-11.

[80], *Höganäs iron and steels powders for sintered components*. Höganäs, (1998).

[81], *Metallography, Höganäs Handbook for sintered components*. Höganäs, (1999).

[82], B. Lacey and C.R. Brooks, "Microstructural Analysis of a Welded Damacus Knife Blade Billet", *Mater. Char.* **29** (1992), 243-248.

[83], M. Durand-Charre1, F. Roussel-Dherbey et S. Coindeau, "Les aciers damassés décryptés", *Revue de Métallurgie* **107**, 131–143 (2010).

[84], J. Maréchal, "La nitruration du fer était utilisée par les anciens", *Métaux,* **391** mars 1958, 133-137.

[85], J.D. Verhoeven, A.H. Pendray, and W.E. Dauksch, "The Continuing Study of Damacus Steel: Bars from the Alwar Armory" *JOM,* Sept. 2004, 17-20.

[86], O.D. Sherby and J. Wadsworth, "Damascus Steel" *Scientific American*, **252** (1985), 112-120.

[87], O.D. Sherby, T Oyama, D.W. Kulm, B. Walser and J. Wadsworth, "Ultrahigh Carbon Steels" *JOM*, june 1985, 50-56.

[88], O.D. Sherby and J. Wadsworth, "Comments on Damascus Steel, Part III : The Wadsworth-Sherby Mechanism by Verhoeven et al.", *Mater. Char.* **28** (1992), 165-172.

[89], E.M. Taleff, B.L. Bramfitt, C.K. Syn, D.R. Lesuer, J. Wadsworth, and O.D. Sherby, "Processing, structure, and properties of a rolled ultrahigh-carbon steel plate exhibiting a damask pattern", *Materials Characterization* **46** (1), (2001), 11-18.

[90], J.D. Verhoeven, A.H. Pendray, "Studies of Damascus Steel Blades : Part I-Experiments on Reconstructed Blades", *Mater. Char.* **30** (1993), 175-186.

J.D. Verhoeven, A.H. Pendray and P.M. Berge,"Studies of Damascus Steel Blades : Part II-Destruction and Reformation of the pattern", *Mater. Char.* **30** (1993), 187-200.

[91], R. Kesri and M. Durand-Charre, "Metallurgical structure and phase diagram of the Fe-V-C system. Comparison with other systems forming MC carbides", *Materials Science and Technology* , **4**, August 1988, 692-699. The Vanadium Shield award by The Institute of Metals in 1989.

[92], M. R. Barnett, R. Balasubramaniam, Vinod Kumar, Colin MacRae, "Correlation between microstructure and phosphorus segregation in a hypereutectoid Wootz steel", *Journal of Materials Science,* **44**, **9**, (2009) 2192-2197

[93], M.R. Barnett, A. Sullivan, R. Balasubramaniam, "Electron backscattering diffraction analysis of an ancient wootz", *Materials Characterization,* **60**, (2009), 252– 260.

[94], J.D. Verhoeven, and E.D. Gibson, "The Divorced Eutectoid Transformation in Steel", *Metall. and Mater. Trans.* **29A** (1998), 1181-1189.

[95], A. France-Lanord, *Ancient Metals. Structure and Characteristics, Technical Cards*, Roma, ICCROM (1980).

[96], W. Kochmann, M. Reibold, R. Goldberg, W. Hauffe, A. Levin, D. Meyer, T. Stephan, H. Müller, A. Belger, P. Paufler, "Nanowires in ancient Damascus steel ", *Journal of alloys and compounds*, **372**, (2004), L15-L19.

[97], M. Durand-Charre, "Nanoscale Precipitation in Genuine Wootz Steel Blades", *http://archaeology.about.com/ Nanoscale Precipitation In Genuine Wootz Steel Blades*. (2013)

[98], P. Paufler, *http://archaeology.about.com/* "*Identifying Mechanical Structures in Damascus Sabers. Are Carbon Nanotubes Present in Damascus Steel?"* 2013

[99], J. Perttula, "Reproduced wootz Damascus steel", *Scandinavian Journal of Metallurgy,* **30**, (2001), 65-68.

[100], J. D. Verhoeven, "A review of microsegregation induced banding phenomena in steels", *J. Materials Engineering and Performance* **9** (3), (2000), 286-296.

[101], T.F. Majka, D.K. Matlock, and G. Krauss, "Development od Microstructural Banding in Low-Alloy Steel with Simulated Mn Segregation", *Metallurgical and Materials Transactions* **33A**, June 2002, 1627-1637.

[102], G. Krauss, "Solidification, Segregation, and Banding in Carbon and Alloy Steel", *Metallurgical and Materials Transactions* **34B**, december 2003, 781-792.

[103], M. Thompson, M. Ferry and P. A. Manohar, "Simulation of Hot-Band Microstructure of C-Mn Steels during High Speed Cooling", *ISIJ International*, **41**, No. 8 891-899.

[104], J. Navasaitis, A. Selskiene, G. Žaldarys, "The study of trace elements in bloomery iron", *Mater Sci* (Medžiagotyra), **16**.2, (2010), 113–8.

[105] A. Williams, "A metallurgical study of some Viking swords.", *Gladius,* XXIX (2009), 121-184

[106] A. Williams, *The Sword and the Crucible: A History of the Metallurgy of European Swords,* Ed. Brill, Leiden, Boston, (2012)

[107] "Scientists determine Viking trade routes by the metal in their swords", *Phys.Org.* Jan 05, 2009
http://phys.org/news150373962.html#jCp

[107] E. E. Astrup and I. Martens, "A metallurgical study of Viking age swords: Metallograph y and archaeology.", *Gladius,* XXXI (2011), 203-206.

13 Index

A
Al Pendray's dagger 120
austenite
  phase field with alloying elements 148
  phase field with chromium 148
  phase field, see phase diagrams
  transformation 132–138, 182

B
bainite 134, 182
banding
  in hypereutectoid steels 170–180
  in hypoeutectoid steels 189–202
blade
  construction (knife) 57–61
  construction (sword) 22–31, 34–41
bolsters 99, 103
borax 161
bulat 78

C
carbides
  chromium carbides 110, 128
  metastable 140
  secondary 153, 172, 181
cast iron
  decarburizing 10
  gray 130
  nodular 11
  production in China 10
  white 132
CCT curves 137–139
Celts, the Celts 22
chromium
  in steel 99, 100, 111, 123
  incidence upon forge welding 168
  phase diagrams 148–151
  steels for cutlery 151–154
coloration of steels
  blades 115
  principles 154
crucible steel 8, 10, 200
crystal structure 128
curvature of the blade 83, 88, 95, 139
cutting edge
  carburization 24
  design of steels 92
  microstructure 114
  quenching to harden see hamon
  stabbing or slashing weapon 14

D
damascene, damaskeened 43, 53, 66
damask naming 47
Damasteel
  formation of the pattern 112–114
dendrite
  arm spacings 145
  microstructure 144
  modelling growth 145
  ripening 145
  segregation 145
  solidification 143, 147
Des Horn 117
diffusion 129, 146
diffusion length, depth of penetration 129
duplex steel 104
Durand Sonia painting 162

E

engraving
  chemical 68
  for damascening 53
Etruscans 31

F

Fe-C phase diagram 131
forge welding
  cycles in practice 123, 175, 187, 200
  effect of composition 167
Forty steps 14, 81
fulah 78
furnace
  primitive for iron smelting 7

G

gilding 68
grain
  definition 143
graphitizing element 74, 98, 130, 154
gun barrels 69–73

H

Hallstät 15, 22
hamon
  Japanese 88
  modern 94
hardness
  in CCT curves 137
  table of values 133
heat treatments 140
Hot Isostatic Pressing 158
hot shortness 170

I

Injection Molding 157
inlay, see damascene, damaskeened
insert 105
interstitial atom 128
Iranian saber 80
iron
  in Africa 11
  mythical aspects 12
  ores 7
  smelting 7
  telluric and terrestrial 5
  working 23

J

Ji-hada 90, 198

K

katana 85
Kellermann's saber 70
Knickmeyer Hank 105, 108, 109
knifemakers
  Chambriard 114
  Chomillier Alain and Joris 101
  Jay Fisher 99
  Rados Jerry 104
Knights-farmers 56
knives
  coloured 115–118
  Damasteel 114
  in the year thousand 57
  modern wootz blades 121, 123
  pattern welded 101, 104
Kris 74–77

L

La Tene 16, 22
Ladder of the Prophet 14, 81
ledeburite 132, 181
legends
  famous swords 14
  forging Mimung 167
  mystery of wootz 200
  quenching wootz 181
  Saladin 47
Loristan 19
lost technique 52, 81, 85

M

macrostructure of ingots 122, 148
martensite
  completed transformation 134
  microstructure 135, 190
  plates or laths 134
  precipitation in martensite 140
  start Ms, finish Mf 134
  tempering 141
  transformation 133
Masson Sébastien 94

INDEX

metastable carbides 140
meteorite
    decoration of the handle of a knife 117
    mythical aspects 12
    polished section 4
    structure of meteoritic irons 5
    use for weapons in Indonesia 74
mosaic 105, 105–109

N

nickel
    graphitizing 98
    in colored blade 118
    in Indonesia 74
    in meteorites 5
    in steels 99

O

ores 7
orientation of grains or particles
    ESBD 180, 194
oxide
    dissolution 161
    distribution of inclusions 201
    inclusions 60

P

Paladru 55
pattern
    moire pattern visible to the naked eye 172, 198–199
pattern welded
    manufacturing 39
    Merovingian swords 35–42
    steels 98–101
pearlite
    divorced 174, 182
    microstructure 133, 172, 182, 196, 197
Petitjean Matthieu 102
phase
    definition 127
    diagrams 132, 149, 171, 176, 192
phosphorus
    effect on welding 163–166
    hot shortness 170
    in ores 10
Pitaud François 116
Plazen Eric 96

powder metallurgy
    manufacturing damask 110–112
    metallic powders 156
    shaping 157
    sintering 157–158
precipitation
    after tempering 97, 142, 172, 185
    in stainless steel 153
    in the wootz matrix 181, 184, 185
pulad 78

Q

quenching
    evidence of practice 26
    fluid 138
    selectively the sharp edge 93
    temperature 134
quillons 66, 67

R

rapier 66
reactions (eutectic, eutectoid, peritectic) 130
restoration 17
Reverdy Pierre 100, 106, 107

S

Sachse Manfred 91
salmon of sword 24
samurai 84
sandwitch steels 95–96
Schneider Friedrich 103
segregation
    dendritic, definition 145
    resulting in banding 189
shamshir 80
sintered steels 110–111
Sintering 158
sites for archeological remains 15
smelting 7
smiths
    Al Pendray 120, 174
    Des Horn 117, 118
    Knickmeyer Hank 108, 109
    Masson Sébastien 94
    Obach Greg 119
    Petitjean Matthieu 102
    Pitaud François 116

Plazen Eric 96
Rados Jerrry 104
Reverdy Pierre 100, 106, 107
Sachse Manfred 91
Schneider Friedrich 103
Veysseyre Jean Pierre 103
Viallon Henri 167
Vulcain 162
Wirtz Achim 121, 123
Yoshihara Yoshindo 86
softening 140
solid solutions 128
steels
   100Cr6, L3, 52100, A573Gr70 138
   composition 98
   dark and bright 99
   duplex steel 104
   for cutlery 148–154
   kawagane, shingane 88
   stainless, optimization Cr content 148
   ZDP189 151
stress relieving 140, 141
substitutional atom 128
sulfur
   hot shortness 170
swords
   attribute of social status 13
   bent 15
   Celtic 22
   Etruscan 31
   famous swords 14
   Gallic 25–30
   Japanese 83–90
   Loristan 19
   Malay Kriss 73
   Merovingian 18, 33–40
   out of Damascus steel 78–81
   Viking 40, 200

T

tempering martensite 141

U

Ulfberht Viking swords 43, 200

V

vanadium (role of) 175
Verhoeven-Pendray procedure 173, 178, 199
vestiges
   archeological remains 15
   treatment 17
Veysseyre Jean Pierre 103
Viallon Henri 167
Vulcain 162

W

Wadsworth-Sherby procedure 173, 199
weld tracks 21, 25, 26, 29, 34, 57–60, 97, 160, 167, 196, 197
Widmanstätten 5
Wirtz Achim 121, 123
wootz
   cake 169
   carbide alignments 120, 174–179
   reconstituted 119–123
   structure of the matrix 183

Y

Yoshihara 85

Z

ZDP189 steel 151

www.ingramcontent.com/pod-product-compliance
Ingram Content Group UK Ltd.
Pitfield, Milton Keynes, MK11 3LW, UK
UKHW061224180426
11947UKWH00027B/2003